OUT OF THE WORLD.

EDINBURGH:

H. & J. PILLANS, PRINTERS,

12 THISTLE STREET.

Near Dunvegan.

OUT OF THE WORLD;

OR,

LIFE IN ST KILDA.

BY

J. SANDS.

SECOND EDITION, CORRECTED AND ENLARGED,

WITH

ILLUSTRATIONS, ETCHED ON COPPER, BY THE AUTHOR.

" She lay like some unkenned-of isle,
Beside New Holland,
Or where wild meeting oceans boil,
Besouth Magellan."
—BURNS.

EDINBURGH:
MACLACHLAN & STEWART.
LONDON: SIMPKIN, MARSHALL, & CO.

MDCCCLXXVIII.

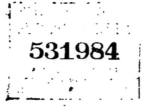

PREFACE.

FAR amid the melancholy main, forty-nine miles west from Obe, in the Sound of Harris, and forty-three from Shillay, in the Outer Hebrides, there is a group of rocky islands, evidently of volcanic origin, the largest of which, called Hirta by the natives and St Kilda by strangers, is inhabited by a small community, who have all Highland names, and speak Gaelic only. This island, which is about three miles long by two broad, is faced on the north-east and south-west by enormous precipices, which rise like walls out of deep water. There is no regular communication between St Kilda and any other part of the world, except by a smack sent by the proprietor twice a year—namely in Summer and Autumn—to collect the rents, furnish supplies, and carry away the produce. Some bold yachtsman generally pays the island a hurried visit about the end of Summer, but as the anchorage is dangerous, he seldom or never remains for more than an hour or two. Few of the inhabitants have been farther than three miles from home; yet although they lead such secluded lives, they are by no means savages. They can all read the Gaelic Testament—they are sober, industrious, decent, and courteous. Although the bravest of the brave when engaged in their dreadful trade of fowling, they are peaceable amongst each other, and never fight. Crime is unknown.

It will be admitted that one may travel far and see nothing more worthy of observation than this little island and its primitive community. In this belief I paid a visit to St Kilda in the Summer of 1875, and remained for seven weeks there. I visited it again, for a purpose to be afterwards explained, in 1876, and resided for eight months. I shall now attempt to describe what I saw, in the hope that it may interest those who have no leisure, inclination, or opportunity for seeing the place for themselves.

Although St Kilda is a small field, yet to give a complete account of it would demand a variety of faculties and acquirements which are seldom or never to be found in one individual. A knowledge of geology, archæology, ornithology, ichthyology, and botany would be requisite. A thorough knowledge of Gaelic, as well as of the Norse tongues, with all the changes they have undergone, would be essential, so that the writer might be qualified to draw philological inferences as to the history of the people. An intimate acquaintance with the popular tales of all nations would also be of great service. The author would, moreover, require to be a good draughtsman; and last, not least, be capable of expressing himself in correct and attractive English. In short, more than the accomplishments which were considered requisite for the poet in Rasselas would find scope in describing the little island of St Kilda. To the possession of these, or of any one of them, I make no pretensions. My sole excuse for presuming to describe the island is based on my having resided on it for a much longer period than any stranger has done for generations, with the exception of the minister, and perhaps some of the former factors. None of those who have written articles on St Kilda have been there for more than a few hours, and have consequently had no opportunity, however gifted, of

observing and understanding the inhabitants or their
sequestered abode. What information these authors
are able to supply must of necessity be borrowed.

To give a picture, drawn from the life, of the little
community so peculiarly situated, is my principal
aim, and the rest of my experience is only sketched
in as a background. Human nature, with all its
failings, is more interesting in my eyes than Nature.
Man is the statue, and everything else but the
pedestal.

CONTENTS.

———

FIRST VISIT, 1875.

CHAPTER I.

CHAPTER II.

CHAPTER III.

CHAPTER IV.

b

SECOND VISIT, 1876-77.

LIST OF PLATES.

OUT OF THE WORLD.

FIRST VISIT, 1875.

CHAPTER I.

Dunvegan—Recommendation to the Inn—Passage to St Kilda
—First impressions — A child's funeral — Visit from the
fair — The church — I get a sermon to myself, but do not
feel flattered.

N the 20th May 1875 I arrived in Dunvegan,
Isle of Skye, having hurried thither post-
haste and regardless of expense by rail and
mail gig from Edinburgh to catch the smack that
was to sail, as I had been advised, next day for St
Kilda, and on board of which I had been promised a
passage. But having reached Dunvegan, I was
informed that the smack would not sail until the
wind changed and the weather became settled. I
also learned from other sources that wind and weather
were of no consequence, as the smack was not there
to profit by a change should it occur, but was engaged
in a cruise somewhere else.

A

" You had better make up your mind," said a new acquaintance, "to stay in the inn for a fortnight, as the smack cannot be in Dunvegan before that time."

" A fortnight!" echoed another, seeing disappointment in my face, " more likely six weeks; and if the weather be stormy, you may not get to St Kilda this year."

This was rather discouraging; but disliking to retreat, I took up my quarters in the inn, and I have much pleasure in recommending that house to any stranger obliged to live in Dunvegan, because there are no other lodgings to be had in the village or vicinity. The charges are strictly moderate, considering the accommodation, and the latter very good, —I mean for a place like Dunvegan,—and the goat-eyed landlord, who talks both English and Gaelic as if an ear of barley had stuck in his throat, is polite and courteous to your face. A rumour that fever raged on the premises caused them to be shunned by all except a few cattle-coupers, who would have had their whisky although Death himself had acted as waiter, and had brought in the stoups and glasses on a coffin lid.

This agreeable hotel is situated at a convenient distance from the churchyard, and thither I sauntered on the evening of the day of my arrival, in search of the antique or curious. There is a church in the middle of the graveyard, not old, although a roofless ruin. Whilst I was meditating amongst the tombs two women entered the ground. One was old, lean, pale, and haggard. She wore a black handkerchief over her white cap. The other was middle-aged, with a dark Celtic face, broad cheek bones, aquiline nose, and black eyes flashing under level brows. She had a dark plaid over her head, and carried a stick in her hand—picturesque figures both. The old woman tottered to a grave and knelt at the head of it, and in an agony of grief patted and embraced the grassy hillock with her skinny hands, muttering meanwhile some words, broken by convulsive sobs, in Gaelic. The younger woman stood composed and erect at the side of the grave, and waited silently until her companion had given vent to her affliction. She then explained to me in imperfect English that the old woman was mourning for her son, a fine young man, who died at the age of twenty-one. Whilst I stood looking at them, the sunbeams breaking like a fan through the murky clouds, I thought that one might spend a lifetime in the lowlands and not find such a subject for a picture. The *Gael*, though more cautious and reserved in his ordinary demeanour than the *Gall,** is more demonstrative in his grief. When

---

* *Gall* (plural *goill*) is the Gaelic for a lowland Scotchman, whom the Highlanders, I believe, never call a *Sassunach*, as

his heart is pierced by some deep emotion he flings off all disguise, as he used to fling off his clothes when he rushed into battle, and his soul stands naked before you.

The modern fashion of levelling the ground and decking it with flowers has not yet been adopted in Dunvegan, and I hope never will be. There the earth is piled up in bold relief, and I like the old style best. It gives assurance to the sight that the poor.inhabitant below is actually there. Why should we try to efface all mark of him? The great majority of the graves have rough stones without any inscriptions at the head and foot.

I entered the ruined church, and whilst I stood staring at some genteel skulls that were lying on the turf beside a handsome tombstone, the two women followed me, and the old one slipping through a gap in the iron railing, brougnt one and put it in my hand. I glanced at its form, and wondered, as one always does in similar circumstances, to whom it had belonged. "Whose skull is this?" From its whiteness I inferred that he must have been a person of quality, too exalted to be interred in common earth.

has been erroneously asserted, this word being exclusively applied to an Englishman. This error originated, I think, with Sir Walter Scott, and has often been repeated since. In this manner history gets corrupted. The St Kildians call all strangers *goill.* If I am wrong let a *Gael* correct me.

" Cha'n innes do ghnuis
A nise co thu
Ma's righ no ma's duic thu fein."

I felt less surprised at the innovations that had
occurred in the Isle of Skye within a century than at
so many things remaining unchanged. The huts of
the peasantry are still exactly the same as they were
in Dr Johnson's time; and here in this churchyard I
had an example of the indifference of the people to
the relics of mortality which struck such horror to
his soul. On mentioning the fact to some acquaint-
ances in Dunvegan, they explained that the bones on
the floor of the church were old ones that had fallen
out of a hole in the wall.

Dunvegan was at one time the great fountain-head
of bagpipe music. No piper was considered finished
unless he had been for several years at the college of
the M'Crimmens. But although the fame of the latter
is still fresh, the *piobrachd* of Donald has been ex-
tinguished by the Psalms of David. The *piob-mhor*
appears even to be regarded as profane in Dunvegan,
or as forbidden fruit—pity 'tis, 'tis true.

I paid a visit to the old castle, which " hath a
pleasant seat" on the banks of the Loch, and is
now adorned by wood on the land side; but I was
shocked to find that the interesting structure had
been damned by improvements. I felt more interest

in an ancient *burg* that stands about two miles distant.

I went into many of the cottages in the neighbour-hood, and in looking back, feel an increased admira-tion for the hospitable character of the people. If I knocked at a door I was always invited in, although they must have known that I might be the bearer of infection, of which all the Highlanders have an uncommon dread. I saw some of the women standing in a dense cloud of peat reek and knitting stockings. The patterns were of their own designing, and I question if any artist in Edinburgh could improve on them. I found both sexes gifted with instinctive politeness. I would have enjoyed my stay in Skye exceedingly, if I had not meant to go further; but having one foot on sea and one on shore I grew impatient to be off, and at length the smack came into the Loch, and on the 1st of June I went on board of her.

The *Janet* (for that was her name) was a very small craft of perhaps 30 tons, if so much, and was laden with stores, such as oatmeal, leather, salt, ropes, and lines, groceries, and dry goods. She was chartered by the proprietor, and the factor was to go out in her. Her crew consisted of the captain and a man. She was to have a boat in tow—a present from Mr Young, of Wemyss Bay, to the St Kildians;

and a man was engaged to sail and steer the boat, with the assistance of the factor's servant, so as to make her as little of a drag on the smack as possible.

At four in the afternoon we left the pier and ran down the Loch with a fine breeze. On the way a cliff was pointed out to me where one of the M'Crimmens (hereditary pipers to MacLeod of Mac-Leod) used to practise his immortal *piobrachd,* and I lifted my bonnet as we passed the sacred spot. At the mouth of the Loch, a boat, filled with weather-beaten fishermen, clad in ochre-coloured oilskins, danced on the water. They flung four ling on board, and bawled a few words of Gaelic as the smack luffed up for a moment.

The long line of the outer Hebrides appeared ahead in aerial hues across the sparkling azure waves ; but the wind fell off, and gloaming had begun before we reached the Sound of Harris. We cast anchor for the night in what seemed a secure and sheltered spot ; but it was said to be quite the reverse in stormy weather.

" Now fades the glimmering landscape from the sight,
And all the air a solemn stillness holds."

The hills were wrapped in the soft, dark shadows of evening, and the thatched roofs of huts, and smoke of peat fires, could just be distinguished through the gloom. The sky and the water alone remained bright. A boat with a dark lug-sail glided close past the smack, the ripple at her bluff bow sounding loudly in the still air. A solitary figure stood at the tiller, the whites of his eyes just visible under the shady brim of his sou'-wester. He passed with a steady stare, but without uttering a word. But the Sound of Harris is not always so silent and peaceful as I found it that evening. The passage is said to be beset with all sorts of dangers—with currents, and bores, and islets, and rocks, so that great experience is necessary to navigate it with safety. Our skipper, however, was familiar with all its difficulties, and had the weather been more unfavourable, would have been equal to the occasion.

I had been well premonished that the smack had no accommodation for passengers, and I did not expect to find the cabin of the *Janet* like that of a Cunard steamer. It was the shape of a quadrant box or a cocked hat, but a little larger. The deck was so low, that when sitting on the locker I was obliged, like the people of Laputa, to keep my head inclined either to the right or to the left. There was a bunk on each side.

The factor being unable to sleep in his berth had had the locker widened for a bed, which occupied about half the cabin. The captain gave up his

Isle of Boreray.

bunk to me, which was not the less kind, although I found it too close to sleep in. I have often wondered whether sailors breathe with lungs the same as human beings, and if so, why it is that they can do without oxygen for four hours at a time. I never happened to fall upon the subject in any book on physiology that I have read. They are so buffeted by the wind that they seem to regard the air as an enemy whom they should take every pains to exclude when they go to rest. The bunk would have been bad enough had it been left open in front, but boards had been nailed on at the foot and head, an aperture being left in the centre through which one had to wriggle like a weasel. Whichever way one turned a board touched the nose. Ten minutes in this catacomb was a long enough time for me, and I was obliged to crawl out in a half-suffocated state, and thrust my head through the hatch for a gulp of fresh air. I then lay on the locker, the edge of it being on my spine, and with one foot up and the other down, enjoyed a sort of dog sleep until morning, when on going on deck I found the smack again under weigh, and Harris some miles astern, whilst St Kilda could be distinguished like a little cloud on the western horizon. But the wind again grew so light that we were still a considerable distance from it at sunset, although

the outlines of it and its satellites were to be seen
clearly defined in purple tints against the golden
sky. I slept for another night on the locker.

Early on the morning of the 3rd, it being a dead
calm, four of us took the presentation boat, and towed
the smack into the bay of St Kilda. I found the
row longer than I anticipated, having been deceived
as to the distance by the magnitude of the cliffs—
an optical delusion I frequently experienced afterwards. The bay is protected in front by a long
craggy island called the Dun ; but it is open to
the south-east, and when the wind blows from that
point, vessels at anchor are in a dangerous position.
Consequently the few adventurous yachts that come
into the bay remain only for a short time. But
although an insecure harbour, it is the only one in
the group of islands.

The village (there is only one) is built on a
comparatively level piece of ground at the foot
of steep and lofty hills. This ground is all enclosed
by a strong stone wall, and contains, as I guess, about
twenty acres. It is subdivided into plots, in which
barley, oats, and potatoes are cultivated. The
village consists of sixteen cottages, and three houses
of a rather better sort, and a church. One of the
best houses is used as a store, another is the manse,
and the third is the house occupied by the factor

Shore near the landing place.

during his brief visits to the island; for he only comes once a year, and remains for about three days to collect the rents which are paid in kind. This unfurnished house he had kindly given me the use of during my uncertain stay, for I had come with the resolution to see the island and its people at my leisure, and to trust to chance for the means of getting off again.

For some time after we had dropped anchor not a soul was to be seen in the village; but at length we could see female forms ascending one of the hills on their way, as I afterwards learned, to the glen to milk their cows and ewes, and a number of men began to gather at the landing-place. The figure of Mr Mackay, the minister, could also be distinguished sauntering from the manse.

After a time a boat came out to the smack, and I was rowed on shore, and found myself standing on the slippery rocks in the midst of the natives. All grasped my hand with a hearty "*Cia mar tha thu?*" They were ruddy, healthy-looking men, dressed in trousers and vests of rough blue cloth, and shirt sleeves of blanketing. They were all barefooted; some of the faces were picturesque, others handsome. They begged for tobacco, but when I assured them I had none to spare, they never troubled me for it again. They were all bawling in Gaelic, with robustious voices, and all at one time.

I was then introduced to Mr Mackay, who, sitting down on a stone, and waving his hand to another for me to sit on, told me he had lived for eight years on the island, and had not seen a newspaper for as many months. He still manifested an interest in the outer world, and asked eagerly for Mr Gladstone, Bismarck, and Dr Begg—all enemies of the Pope. In a small field between the manse and the factor's house is a well of excellent water, which never fails.

The churchyard is close behind the village, is very small, and of an elliptical form. It is fenced with a stone wall, and the gate is kept carefully closed. On the 5th I attended the funeral of a child. At this time I was not aware that the infants of St Kilda were subject to a distemper which cuts off the great majority when they are a few days old. On this occasion I saw two little boxes disinterred to make room for the new one. I then went with the men who had dug the grave to the house of mourning, where about twenty women were squatting on the floor, the bright coloured handkerchiefs on their heads gleaming through the smoke of the dingy apartment. An elderly man offered up a long prayer in Gaelic, whilst the women chanted a monotonous tune in a low tone. The father of the child then put the small coffin under his arm, and we all followed in single file up the rough pathway that serves for a street to the graveyard. I heard that this man had lost eight infants.

A day or two after my arrival, Mr Mackay came to me in a state of excitement, saying that the people had been told I had been living in a house where fever was raging, and he wanted to know if it was true. I admitted that I believed there had been fever in the inn, but had heard that those who had been ill had now recovered. I had no personal knowledge on the subject. The minister then returned with my explanation to a group of the natives, who were clamouring and gesticulating in evident agitation about a hundred yards off; and as I never heard any more of the matter, I suppose he had succeeded in assuring them that there was no danger. I had been told on board the smack, although without any refer-ence to my own case, that the St Kildians have had, ever since the island was scourged by the small-pox in 1730, a mortal terror of infection. I have no doubt, but for the intercession of the minister, that I would have been expelled from St Kilda before I had had time to see it properly; and I suspect that some one in the vessel had disclosed the above rumour in order to bring about that consummation.

On the 6th the *Janet* sailed, and I was left alone in the house, with no means of escaping from the island if I grew tired of it. In the evening, about twenty women in a body paid me a visit, each bringing a burden of turf in her plaid, which they piled up in a corner of the room, as a gift. After standing for a few minutes with pleasant smiles on their sunny faces, they departed with a kindly "*Feasgar math libh!*" I was subsequently honoured with frequent calls from the fair sex, and, like misfortunes, they never came singly, but in crowds. I had still more frequent visits from the men, who also came altogether if they came at all. Their visits were kindly meant; but as they talked, or rather bawled, with voices like Stentor, and all at one time, I was at first driven almost distracted, and sometimes seized the pipes (the largest size) and blew up to drown the din. Their good nature, however, rendered this of no avail, as they would wait until I had finished, and then resume the altercation as if it had never been interrupted. But after a time their visits suddenly ceased, from which I inferred that my jocular grumblings had been communicated to them by the minister. They, however, remained as friendly as ever.

One of my neighbours agreed for a consideration to bring me water and milk, and to sweep up my hearth (a bundle of solan geese wings serving for a

broom) once a day; and having been accustomed in
my time to roughing it under more adverse circumstances, I had no fear about being able to cook for
myself in a plain way. For recreation I had brought
with me a set of Highland pipes and a keyed flute,
and I had no books but the Gaelic Testament.

On the 7th I attended the Free Church, to which
denomination all the people belong. The building
itself is a very plain—not to say ugly—structure.
It belonged originally to the Established Church,
but the latter allowed it to be taken possession of by
the Frees, as the swallows allow their nests to be
taken from them by the sparrows. A bell saved
from a wreck called the people together, and all the
population seemed to attend. The men wore rough
jackets in addition to their week-day attire, and the
women dark plaids; but all the females were barefoot. I had never seen a more earnest and attentive
audience. Every eye was fastened on the minister
throughout the whole discourse. He is one indeed of
whom it may be said that he commands attention,
for if any woman, say, happens to drop asleep, she is
immediately aroused with a "*Lachlan, dùisg a bhean;
cha bhi cadal an Ifrinn*—Lachlan, awaken your
wife; she wont sleep much in hell, I think!" which
causes Lachlan to stick his elbow into his wife's ribs
immediately. The minister then indulges in a low

chuckle, mingled with coughing and inarticulate sounds, and waits patiently until the woman is thoroughly aroused.

At the conclusion of the morning service, Mr Mackay invited me to lunch, and apprised me that he intended to deliver an address in English in the afternoon for my sole edification. Alarmed at this proposal, I remonstrated, when he replied, " But you don't understand me when I preach in Gaelic." " Oh, never mind," I answered, " I understand as much as I want to." " Say that, say that, I believe you are speaking the truth there." However, he appeared, although with reluctance, to accede to my desire, and I walked into the trap. Nevertheless, after a few preliminaries in the vernacular, like smoke to conceal his movements, he suddenly opened a battery in English, and blazed away for an hour. As I was the only one present who understood a word of that language, by Jove ! the attack seemed personal. I did not know whether to feel angry or amused at the situation. I endured it, but I could not help thinking that it was ungenerous in him to take advantage of a poor sheep that had entered confidingly into his fold, and worry him in that way. Some of the congregation looked at me with weary glances, as the spectators at a banquet look from the gallery upon the lucky fellows who are feasting below. I may mention that my motive in going to the church (although no apology is needed for that, I suppose) was to try and pick up a little Gaelic, the study of which I had only commenced a few months previously, and the want of which I began to regret every day.

B

CHAPTER II.

History of St Kilda—Population—*Tetanus infantum*—The
people, appearance, character, education, religion, customs,
dress, etc.

THE history of St Kilda, like our own, is lost
in antiquity. It would be interesting to
learn what brought the first settlers to this
lonely and desert isle, and to ascertain the date of
their immigration. Were they castaways? or were
they attracted to these rocks by the multitudes of
sea-fowl with which they are frequented, and whose
flesh and eggs still form important articles of diet?
Or were they fugitives, driven from their homes by
powerful enemies, and glad to fly with their families
to this remote island, where they might hope to live
in peace, or at least have an opportunity of defending
themselves? for the fort on the Dun testifies that
even here they did not expect to live unmolested.
The old huts still remaining are primitive enough,
but the first dwellings were probably underground.
One of these subterranean houses still exists in the

village, and there are numerous caves which, no doubt, were used as places of refuge in times of danger.

It would be interesting to know how the first proprietor obtained possession. " It is, I believe," says Macaulay, " no easy matter to trace out the name, nor of course the history, of *Hirta*, with any degree of certainty, beyond the fourteenth century.

" In a charter granted within that period by John, Lord of the Isles, to his son Reginald, and confirmed by King Robert the Second, St Kilda, under the name of Hirt, was made over, together with many other places, to the said Reginald. How at the end of two or three generations the property of this isle was transferred from the successors of Reginald, the predecessors of Clan Ronald, to the family of Sleat, now represented by Sir James M'Donald, and how in process of time it fell into the hands of the clan that now possesses it, is a useless inquiry. And were the question of greater importance, so contradictory are the accounts given, and so slender are the historical evidences on every side, that any judicious person will choose to leave that matter undetermined. At this time (1759) the proprietor is Norman MacLeod of MacLeod ; and his ancestors have possessed it for at least two hundred years back."

The St Kildians have a dim tradition that their

ancestors came from Uist. They also believe that the island has been depopulated more than once and planted anew. They tell an incredible tale, which, however, they firmly believe in, about two men who, at some remote and indefinite period, decoyed all the rest of the people into the church, and there suffocated them. From the appearance of the people, and from the Norwegian words with which, according to Macaulay, their Gaelic is corrupted, I conjecture the present inhabitants have some Scandinavian blood in their veins. The Norwegians (who possessed all the Hebrides for several centuries) have apparently left their mark on the present inhabitants of St Kilda. "The people of this island," says Macaulay, "have a tradition that one Macquin, an Irish rover, was the first person who settled himself and a colony of his countrymen in their land."

The population has much decreased since Martin's visit in 1697, when it numbered about 180. When the Rev. Kenneth Macaulay (although I do not now consider him a reliable authority) was there in 1759, the population had decreased to 88. It was reduced to 72 in 1875, of whom 29 were males and 43 females, but increased to 75 in 1876. Small-pox visited the island about 1730, and made fearful havoc in the little community. The infection was said to have been brought from Harris. Of twenty-one families only four adults remained, and these had the burden of twenty-six orphans to support. " Before the distemper was propagated three men and eight boys were sent into one of the islands with a design of catching solan geese for the benefit of the whole community. An universal confusion and mortality ensuing at home, they continued there from the middle of August till about the middle of May in the following year. The boat in which these men had been wafted over into that island was brought back to Hirta before the distemper became epidemical. Had they been at home with the rest, it is more than probable that their fate had been the same as their friends."*

Thirty-five persons left for Australia about twenty years ago, most of whom died of ship-fever on the passage. The population was further reduced by the loss of seven men and a woman, who were drowned in attempting to reach Harris in 1864.

But the decrease in the population of St Kilda is chiefly owing to a disease to which the infants are subject, and which cuts off a large proportion a few days after birth. Medical men call it *tetanus infantum*, and the Irish "nine day fits." Doctors differ as to the cause; some say that it arises from the mothers living on sea-fowl; others to deterioration of

* Macaulay, p. 198.

the system from long-continued intermarriage; some say the infant is smothered with peat-reek; whilst some assert that a small operation necessary at birth is improperly performed; and others declare that the infant is killed by improper feeding. Having seen a case or two, and having had an opportunity of making inquiries on the spot, perhaps I may, without presumption, venture to say that I think the disease is attributable to the latter cause. Women, who are about to be mothers, go when they have an opportunity to Harris to be confined, so as to escape the curse that seems to hang over their progeny in St Kilda. This scourge must be all the more severely felt from the remarkable fondness which these poor people have for the young. The few children on the island are idolised. I have frequently sat in a corner and observed a group around an infant, every head, old and young, craned towards it, every face beamed with smiles of delight. This disease has prevailed for upwards of a century at least, and it seems to me that those who have been entrusted with the management of this secluded island ought to make a determined effort to discover the cause, and to put a stop to it. I have heard more than one pious gentleman suggest that this distemper was probably a wise provision of Providence for preventing a redundant population on a rock where food was limited. But I cannot believe that healthy, well-made infants were sent into the world to die a few days after birth. If the island became too small for the population, the excess might be removed to another place; strong men and women ought to be valuable. It is wonderful with what ready submission some good Christians can kiss the chastening rod when it is only to be laid upon their neighbours' backs.

If the children escape the *tetanus infantum* they grow up into fine boys and girls, and strong men and women. The average height of the men is 5 feet 6 inches. The tallest man is 5 feet 9 inches, the shortest 4 feet 10½. I measured twenty-one male adults. Both sexes look strong and healthy. They have ruddy cheeks, remarkably clear eyes, and teeth like new ivory. Their limbs are as hard as boxwood, and their whole frames capable of severe and long-continued exertion. Although they have inter-married possibly for a thousand years, and at all events for several centuries, none of the pernicious effects that one has been taught to expect seems to have resulted. Only one man has been afflicted with imbecility, and that not of the worst sort. He lives in one of the old hovels, and manages to support himself by cultivating a bit of ground, by keeping a few sheep, and by catching birds when the others allow him. He is peaceable and even polite when let alone, but a terror to all when enraged. He is old, and not quite fairly used I thought. He attends the church regularly, and shakes his head at the sins of his neighbours as sensibly as any intelligent person could do. He has a sister afflicted with epilepsy. One woman has a deformed foot, who has a daughter with a deformed chest. There is an old man blind from cataract. All the rest of the people (except one old man crippled by accident) are sound in mind and body, and enjoy an immunity from the diseases that afflict humanity in places of greater material civilisation. There is a large proportion of old people in the community, although none have attained a great age. It is remarkable that only two of the natives are bald-headed, and that the hair of many people advanced in years has retained its original colour. Many of the old have also good teeth, although tooth brushes are unknown. The generality have strong sight.

Complexions vary. Some have hair like flax, and others like the raven's wing. Some have fair skins and blue eyes, and others dark skins and brown or black eyes. Their minds are wonderfully acute considering their contracted experience, which, in the most of cases, has been confined to the rock on which they have been born. Some have been as far as Lewis, Harris, and Uist, and even Skye, and surprise

Natives of St Kilda.

the others with an account of the wonders they have
seen in these distant places,—a man with a wooden
leg having apparently created the greatest interest.
One man told me he had once seen a drunk sailor,
and mimicked his walk and gestures as if the spectacle
was one not often to be seen and never to be forgotten.
A woman then capped the story by declaring, with
open eyes and mouth, that she had seen a drunken
doctor who had been sent to the island to vaccinate
the children, but had to be carried on board the
vessel that had brought him like a corpse.

All beyond their little rocky home is darkness,
doubt, and dread—incomprehensible to us. In
ordinary moods they are cheerful and composed, but
a very small occurrence throws them all into a state
of excitement.

Great care has been taken of their religious education, and all can read the Gaelic Bible with more or
less facility. Some can repeat entire chapters by
heart, and you can scarcely refer to a passage which
they are unable to find at a moment's notice.
Their secular education has been less cared for, and
indeed the difficulties of imparting it are very great,
as these poor prisoners have no inducements to learn.
Why should they trouble themselves learning to
write when they are cut off from correspondence
with all the world? A few, however, can write,

although painfully, in the vernacular. Why should they learn English, when, in the course of a year, they may only see two or three strangers who use that language, and then only for an hour or two? *Beurla*—English—they say, "Why, it is like the cackle of a fulmar." Geography they seem to feel more interest in, and some of the men asked me, " Where is California ? where is Australia ? " Having no maps I was obliged to indicate the forms and relative positions of these countries (where St Kildians have settled) by tearing bits of paper and placing them on the ground. One man, who is reputed a good scholar, was obliged to mutter *units, tens, hundreds, thousands*, before he ventured to read 1875. They have heard of the telegraph, and speak of Mrs Somebody, the post-mistress in Obe, Harris, as the most extraordinary genius that ever lived, because she is able to work it. The minister acts as school-master, and teaches, in a desultory way, the six or seven children who are of the proper age for instruc-tion. But the man is old and has a disordered liver, and perhaps does not see the importance of giving the young a better secular education. I should be very sorry to see the Gaelic fall into disuse in St Kilda, of which there is very little fear; but I think it would be well if a few were taught English, so that they might be able to read newspapers and books

for the benefit of the rest, and speak to strangers
without an interpreter. The minister, and a woman
from Ross-shire, who is married to a native, are the
only persons who can talk English, and they can
give any representation as to the opinions of the
people they choose. This accomplishment gives them
greater influence than either of them is entitled to
possess.

The St Kildians are distinguished by six names,
viz., Gillies, Fergusson, M'Donald, M'Queen, M'Kinnon,
and M'Crimmen. But in many cases, for the sake of
distinction, an adjective is added to the Christian
name, or the person is referred to as the son of his
father. For instance, Findlay Gillies is called Findlay,
the son of Rory, or Findlay, the son of Tormad.
Donald M'Queen is called Donull Og—young Donald,
and Callum M'Donald is called Callum Beag—little
Malcolm. These names continue in use long after
they have ceased to be appropriate, young Donald
being now one of the oldest men in the island, and
little Malcolm a young man of good stature.

The St Kildians have, and seem always to have
had, a more than ordinary development of the religious
instinct. Altars to pagan gods were erected on the
hill-tops, and continued to be used not only throughout Catholic times, but even when Presbyterianism
was firmly established, and a minister residing in the

island.   The three chapels that formerly existed in *Hirta* prove that religion was not neglected before the Reformation, although Buchanan says that in his time the St Kildians were ignorant of all the arts, and especially of religion.

In 1697 every man seems to have been his own minister, and a clergyman was only sent occasionally to ratify the ceremonies that had been previously performed.   At this period a native called Ruari Mòr pretended that John the Baptist had appeared to him, and given him a commission to rule over the commonwealth.   John and he seem to have been on the most intimate terms, and had frequent interviews. By dint of cunning, and a natural power of persuasion, he obtained general credence, and thereby tyrannised over the men and debauched the weaker women. Those who were too virtuous to obey his "earthly and abhorred commands" were subjected to the most ignominious punishments,—such as being compelled to stand, in a state of nudity, under a waterfall in the presence of all the people.   He finally made a confession of his imposition before the Presbytery of Skye.   His surname is now forgotten in St Kilda, but Donull Og can still repeat (to the amusement of the others) long passages from his sermons which have been reported orally from father to son.

The first resident Protestant minister was one

Buchan, who wrote a book about St Kilda, or more strictly speaking, printed Martin's book with his own name on the title page. He is said to have introduced the use of letters, and deserves greater praise for that than for having published Martin's book. The St Kildians say he was killed by a bull that he was endeavouring to show the people there was no danger in approaching ; but this does not seem to be true.

Now-a-days the whole population are members of the Free Church. They attend Divine service three times every Sunday, and hold a prayer meeting (which is conducted by the elders) every Wednesday evening. The Sabbath is indeed a day of intolerable gloom. At the clink of the bell the whole flock hurry to the church, with sorrowful looks, and eyes bent upon the ground. It is considered sinful to look to the right hand or to the left. They do not appear like good people going to listen to glad tidings of great joy, but like a troop of the damned whom Satan is driving to the bottomless pit. Surely this is not the proper deportment for good Christians —surely religion, with its promises of remission of sins and [everlasting life beyond the grave, should make true believers more cheerful and not more miserable than benighted heathen, who have no such consolations. Instead of assuming this dejected behaviour, it appears to me that real Christians should march with heads erect, with eyes beaming with exultation, and scorning to look upon the vile earth, and with glad voices singing, " O death, where is thy sting ? O grave, where is thy victory ? "

No one speaks above a whisper or visits another on the seventh day. For myself, I felt like an owl in the desert, and was fain to steal out in the dusk, and stretch my limbs with three steps and a turn before my domicile ; for a long walk, or rather a climb, for there are no walks, was evidently regarded as the height of iniquity, and it was not my cue to offend the people.

Besides the meeting which is held every Wednesday, a meeting is held on the first Tuesday of every month, to return thanks for the preservation of Captain Otter of H.M.S. *Porcupine*, whose vessel was nearly lost on the island some years ago. This was instituted at the request of the now deceased Captain, who brought them supplies in a season of dearth, and who attempted a number of improvements which have proved abortive. This meeting, it seems to me, savours of Popery.

The poor St Kildians contributed, as far as I can understand from the printed report, £20 last year (1874) to the Sustentation Fund of the Free Church, which must have cost them an enormous effort, and, coupled with the unprofitable way in which their trade is conducted, reminds one of the passage in Scripture: "That which the palmerworm hath left hath the locust eaten" (Joel i. 4).

Family worship is held in every house morning and evening; and when parties of men and women reside in the other islands they "make their worship," as they phrase it, just as they do at home. Every meal is preceded by a grace, nor will they take a drink of milk or water without uncovering the head.

. In the olden time the St Kildians tempered their religious exercises and their dangerous and laborious employments with a little recreation. They ran races on their small but fiery ponies on the patch of sand that is cast up by the waves every summer in front of the village, but which is washed away by the rolling billows every winter. They played *camanachd* or shinty on the same spot, flinging off their belted plaids in their joyful excitement, and going at it in their shirts. They were excellent swimmers in those days. The ponies are now extinct, *camanachd* has been given up, and the men never wet their skins unless when they fall by accident into the water. They were fond of music, too, in the good old days, although the Jew's harp was the only instrument they possessed, and to its feeble twang they danced and were gay. Now dancing is unknown. The women when they sat together in the street used to compose rhymes whilst they plied their spindles, but this intellectual amusement has been abandoned.

The men of St Kilda are in the habit of congregating in front of one of the houses almost every morning for the discussion of business. I called this assembly the Parliament, and with a laugh they adopted the name. When the subject is exciting the members talk with loud voices and all at one time. If this system were adopted in the Imperial Parliament, a deal of time might be saved, and the humblest member have a chance of speaking. Some of the men may be seen reclining on the top of the wall of an old hut, others leaning against it, while two or three stride backwards and forwards with their hands in their pockets and their beards in the air, bawling at the full pitch of their powerful voices, and even rising on tip-toe when the debate becomes vehement. You may hear them half a mile off. A stranger would fancy they were about to come to blows ; but nothing is farther from their thoughts. Shall we go to catch solan geese, or ling, or mend the boat? or hunt sheep? are examples of the subjects that occupy the house. Although exceedingly disputatious, they work in perfect harmony when once the question is settled.

Parliament, besides being necessary to the conduct of business, has, I think, a salutary effect on the minds of the people, and helps to keep them cheerful in spite of their isolated position and excessive religious exercises. Man is a gregarious animal, and there are no people more so than the St Kildians, although the herd is small. In work every one follows his neighbour. If one puts a new thatch on his barn, a man is to be seen on the top of every barn in the village. If the voice of praise is heard at one door, all, you may be sure, are engaged in worship, and so on.

The St Kildians are quite as industrious as they are pious. Every family has a croft of ground, which they carefully cultivate, although their method of husbandry admits of improvement. They grow oats, bere, and potatoes, and a very few cabbages and turnips, all of which are planted too thickly. Besides other manure the ground is enriched with the carcasses of puffins, but there is a great waste of this valuable manure, many thousands of these birds being left every year after the plucking season to rot in the islands of Soa and Boreray. The grain is ground into meal by handmills. In the beginning of summer the rocks are scaled, and the neighbouring islands visited for old solan geese and eggs. They fish for ling in summer, and *pluck* their sheep—for

C

shears are unknown. Farmers tell me that the possibility of performing this operation shows the animals have been half-starved in winter. The wool is spun by the women, and woven by the men into cloth and blankets, which, after providing clothes to themselves, are sold to the factor. In August they catch the young fulmars, and in September the young solan geese. In winter the spinning-wheels and looms are busy from dawn of day until two or three next morning. Their diligence and endurance are astonishing. All the fuel used is carried by the women from all parts of the island. It consists of turf. They can carry enormous burdens down from the tops of the highest and steepest hills. When the boats arrive they hurry to the shore, and lend a powerful and willing hand to drag them over the rocks. They herd and milk their cows and ewes, and make cheese. They are never idle. When other work fails, they ply the spindle or knit stockings.

Both sexes are polite in their own way, and when they fail, it is from inexperience of conventional usages. They never forget to wish one "*Maduin mhath libh*" when they meet one in the morning, and "*Cadal math dhuibh*" when they part at night. The men doff their bonnets with the left hand and hold out the right. All remain seated in Church until the women have made their *exeunt*.

In the virtue, which is next to godliness, I am bound to confess they are deficient; but there has been a great improvement in this respect since the time (quite recent) when the whole year's manure was collected in the hut in which they lived until the floor rose several feet above the proper level. Soap, if it could be got cheap, would no doubt work a greater reform. Several streams of water run through the village, and might be utilised for bathing. Two houses for this purpose could be erected without any cost. The minister, if he must meddle with temporal affairs, ought to inculcate cleanliness, and use all his influence to enforce it; " Point to the bath, and lead the way."

The *breacon an fheili*, or belted plaid, was the dress of the St Kildians when Martin visited the island in 1697, as it was in the Highlands of Scotland at and long after the same period. He says: "Their habit antiently was of sheep-skins, which was wore by several of the inhabitants now living; the men at this day wear a short doublet reaching to their waste, about that a double plait of plad, both ends joined together by the bone of a fulmar; this plad reaches no further than their knees, and is above the haunches girt about with a belt of leather; they wear short caps of the same colour and shape with the capuchins, but shorter, and on Sundays they

wear bonnets; some of late have got breeches, and
they are wide and open at the knees; they wear
cloth stockings, and go without shoes in the summer time; their leather is dressed with the roots of
Tormentil."

Now all the men wear trousers and vests of coarse
blue cloth, with blanket shirts. On Sundays they
wear jackets in addition. Their clothes are made
at home from wool plucked from their own sheep,
which is spun by the women with the ancient
spindle or more modern wheel. The women also
dye the thread, and the men weave it into cloth
and make it into garments for both sexes.

The dress of the women consists of a cotton
handkerchief on the head, which is tied under the
chin; a gown of coarse blue cloth, or blue with a
narrow purple stripe, fastened at the breast with
a pin made from a large fishing hook. The skirt
is tied around the waist, and is girded tightly
above the haunches with a worsted sash of divers
colours, and is worn very short, their muscular
limbs being visible to near the knee. They wear
neither shoes nor stockings in summer and seldom
in winter. They go barefoot even to church, and
on that occasion don a plaid, which is worn square,
and is fastened in front with a copper brooch, like
a small quoit, made by the men from an old penny.

All the women's dresses are made by the men, who also make their *brogan* or shoes, for every female possesses a pair, although she prefers going barefoot; and I am not surprised at this, as the shoes, although substantially made, are as hard as box irons, and not unlike them in shape. The *brogan* are sewed with thongs of sheep-skin. The *brog tiondadh* or turned shoe, so called because it is sewed on the wrong side, and then turned inside out, was in vogue until quite recently, and specimens are still to be seen. It is made to fit either foot. The leather is bought from the factor, but the sheep-skin is tanned by themselves with roots, or bark found under the turf. Conical sheep-skin caps were once common, and are still to be seen.

It is remarkable that in all their works there is no attempt at ornament, in which they differ strikingly from the Highlander, who, when he was at liberty to please his own fancy, decorated his person from top to toe, and who, even now when he has been constrained to adopt the Saxon fashions, abhors everything that is plain and unadorned, and cannot carry a walking-stick until he has carved some fantastic head or knot-work pattern on it. The St Kildians seem deficient in this artistic instinct. The brooches of the women are perfectly plain, and the pins that fasten their gowns mere skewers. No attempt is made to carve in stone or wood, or to decorate cloth or leather.

When any one dies he is interred the same day. As deaths are few and far between, they are proportionately appalling, and occasion a general mourning which lasts for a week, during which no work is done. The grief of the survivors, I have been told, is excessive and unrestrained.

CHAPTER III.

THE people of St Kilda are warmly clothed (thanks to their own industry), and in general well fed. Their diet consists of mutton, sea-fowl eggs and flesh, potatoes, cheese, oatmeal porridge, oat cakes, and milk. They have a prejudice against fish and use it sparingly, alleging that it causes an eruption on the skin. They care little for tea but are fond of sugar, and the women are crazy for sweets. The men are equally fond of tobacco, although they consume little, probably because it is too costly. Every family is provided with a bottle of whisky, but the cork is seldom or never drawn except in case of sickness. They had never seen fruit until I took three apples to the island.

All the inhabitants are congregated in one village, which is built in the form of a crescent facing the bay at the south-east side of the island. The houses

were until lately similar to those in other islands
of the Hebrides, and are said to be warm and
comfortable.　They are at all events picturesque.
The walls are built of loose stones, are about five
feet in thickness, with turf packed in between.　They
are thatched with straw, held down by ropes of the
same material tied to stones.　Some of these old
huts remain, and are now used as barns.　Two are
still occupied; but fourteen new cottages were
erected about fourteen years ago, and have excited
the admiration of visitors, although I think there
is still great reason for improvement as regards
furniture.　These cottages are well built, and are
roofed with zinc.　Each house contains two chairs
and no more, which were sent from Edinburgh.　Very
few of the houses can boast of a table, and each
family sits or squats around the pot when at dinner.
The people make stools of straw ropes.　The
minister is the only person on the island who
possesses or uses a fork.　Almost every house
contains a loom, a spinning-wheel or two, and every
woman has a chest in which to store her MacGregor
shawl (used on great occasions) and other articles of
dress.　None of the walls have been white-washed
since the houses were erected, and they are blackened
with peat smoke.　Bundles of solan geese stomachs
and ropes for the crags hang from the rafters.　Every

Old Dwelling House.

family is well provided with blankets, but they all sleep on loose straw. Some of MacLeod's friends speak in glowing terms of the comfort of these cottages, and quite forget it was owing to the agitation raised by Captain Otter, and to his threatening to start a subscription to provide better dwellings for the St Kildians, that caused the previous proprietor to build the present houses at his own cost. I believe . the rent of each cottage and plot of ground is £2 per annum, which, if the trade were free, would be very moderate.

MacCulloch has compared the form of St Kilda to a leg of mutton. It consists of a number of steep hills, arranged something like the figure 4 as it is written, or, if we include the island called the Dun, like the letter H roughly formed. The highest hill is Connagher, which is 1220 feet above the sea. I have said hills, but in reality they are only the halves of hills — hills to the interior, but cliffs to the sea. Connagher, like the others, is one stupendous cliff almost from the summit to the base. All the rocks are igneous. The space below the bar of the H is the bay, and the space above it Glen Mòr. In this glen is the most extensive pasture for sheep and cattle, and to this secluded spot a large proportion of the women go every day during summer to milk their ewes and cows and herd their flocks. One can hear their shrill voices a mile off as they shout to each other across the glen. On the whole island, as Martin remarks, "there is no sort of trees; no, not the least shrub grows here, nor even a bee seen at any time." But, as he says elsewhere, "The grass is very short, but kindly, producing plenty of milk." Large tracts close to the sea are carpeted thickly with sea-pinks.

Fish abound on the coast, including ling, cod, lithe, halibut, black-mouths, skate, and conger-eels. The latter are used for bait. Two boats, with crews of eight and nine men respectively, went to sea frequently in the evenings during the time I resided on the island this year (1875) to fish for ling with long lines. Each boat would return in the morning with perhaps on an average about thirty-five ling and a few cod, besides other fish. The ling were all salted and stored up for the factor, who pays 7d. each for them. The cod are sold for 3d. each. The men provide their own lines, but are supplied with salt. They themselves keep no note of the number of fish exported, but MacLeod incidentally acknowledged receipt of " 1080 marketable fish."

A number of elderly men were wont to sit on the rocks near the village and angle for bream in the July evenings. For each of these they receive a penny.

The only domestic animals in St Kilda are cattle, sheep, dogs, and cats. The young cattle are all sold to the factor. Every family has a cow, and the minister two. I could not ascertain what number of sheep were in *Hirta* and adjacent islands. The people themselves did not seem to know. Some of them are richer in stock than others. There is a flock of a peculiar breed on the island of Soa, which belongs to the proprietor. The latter run wild, and are hunted down to be plucked in the season. Their wool is of a light brown colour, and makes pretty cloth or stockings without dye. The cattle are very shy of strangers—run off when they see them, and stand staring at a distance. The native breed of dogs has all but become extinct. They are like cream-coloured collies. A lot of mongrels has been imported from Harris. They are trained to herd sheep and catch coulternebs. " The number of horses," says Martin, "exceeds not eighteen, all of a red colour, very low and smooth skinned, being only employed in carrying their turf and corn." In 1759 the number had dwindled to ten, and are described as being of a very diminutive size, but extremely well cast, full of fire and very hardy. There are no horses on the island now, and all the work they did is now performed by the women. Would it not be well to introduce a few ponies again ? Some men over thirty have never seen a horse, and one told me how astonished he felt on seeing one drawing a cart in Harris. He thought at first it was a cow.

The only wild animal is the mouse, of which there are two species, the house and field. I have only seen one of the latter, and it was too distant for inspection. They swarm in Glen Mòr, I was told, and also in the Dun. The people are careful not to carry them to Boreray or Soa, and I heard that a party of men who had gone to Boreray to pluck sheep happened to carry a mouse in their baggage, and dreading that it might be a female who would introduce the breed, they resolved to destroy her at any cost. They had, however, to take down seven *clætyan* before they accomplished their purpose.

That enterprising gentleman, the rat, has not as yet been able to effect a landing, very likely because the few vessels that touch at the island anchor at a distance from the shore.

The mention of *clætyan* (I spell this word phonetically) reminds me that I ought to have described these buildings before now, as they are objects which strike the eye whenever one enters the bay. A clætya, then, is a small house, in which the people used to dry their birds before salt was introduced into the island. It is now used to hold turf, hay, etc. Their number excites surprise. They are crowded at the back of the village, and are to be seen all over St Kilda, up even to the tops of the hills. There are also a number in the other islands. They are primitive looking structures, consisting of two dry-stone walls, so close together that stones laid from side to side are employed to form the roof, which is then covered with turf.

The birds which are of the greatest consequence to the two-legged inhabitants without feathers are the fulmar, puffin, guillemot, and razor-bill. These frequent the cliffs in numbers numberless, and are valuable for their feathers, oil, eggs, and flesh. There are no solan geese in St Kilda, but multitudes inhabit the island of Boreray and adjacent stacks.

The fulmar petrel (*fulmar glacialis*) is like a middling-sized gull. The plumage is light grey on the back, and white on the belly. The bill is strong, and hooked at the point. The nostril looks like a small tube placed on the top of the bill. The fulmar is a humble relation of the albatross. He has a light and airy flight, and seldom flaps a wing, but floats about in graceful curves as if it cost him no effort to keep aloft. He builds his nest on the grassy ledges of the great cliffs, and the female lays one egg, which is white. If this egg is taken away, she lays another. If that is also stolen, she gives up the business in despair. St Kilda is the only part of Great Britain in which the fulmar breeds, although he makes excursions to Shetland. He is said to feed upon the blubber of live whales, porpoises, etc. I opened the stomachs of a number and found nothing but oil. The young birds are caught for the oil, which they try to spit into the faces of the men who catch them. This oil is of a reddish colour, and has at first a very offensive smell—a smell, indeed, which pervades every part of the fulmar, flesh, feathers, and all. He has been well named the skunk of birds. The oil is used in the lamps, and is equal to whale oil. The quantity exported in 1875 was at least 906 St Kilda pints—each pint being equal to 5 English pints. The St Kildians receive one shilling a St Kilda pint for it. The feathers are also sold to the factor, who gives five shillings for a St Kilda stone of them, the stone containing twenty-four pounds. The old fulmars are caught whilst hatching; the young before they can fly. The fulmars (all but a few invalids) leave St Kilda about the end of August, and return about the 10th of November. They are lean and worthless when they come back. The people eat the eggs, and are not fastidious as to their freshness. A chick inside is of no consequence. They fling it away and eat the rest. The carcasses are put into pickle and eaten in winter.

In respect to usefulness the puffin ranks next to

the fulmar. He is about the size of a pigeon. His plumage is black on the back and white on the belly. His triangular bill is deep and narrow, and beautifully marked with vermilion, dark grey, and yellow. He has red webbed feet, which he spreads out behind him as he flies, like a jury rudder. His wings and tail are short and his flight laborious, although he progresses at a rapid rate. People have been provoked by his comical appearance to call him all sorts of names,—a sea-parrot, a coulterneb, a Tammie Norrie, and the St Kildians have borrowed a name for him from the French. The scientific call him *alca artica.* He feeds on fish fry. His note is like the voice of a man in the deepest agony—oh! oh!— the last syllable being long drawn out. He burrows in the ground, and the female lays one egg of a white colour. He is caught in St Kilda by dogs, by snares, and by a noose on the end of a rod. His feathers, which are of the finest quality, are sold to the factor for six shillings a St Kilda stone. A few of the carcasses are roasted and eaten, and taste like a kippered herring, with a slight flavour of the dog-fish. A quantity are used for manure, but a very large proportion are thrown away.

The guillemot (*uria troile*), which the St Kildians call the *Lavy*, is caught when hatching, at which time the female is said to be in the fattest condition.

She is caught by a noose tied to a rod. In spring another method is adopted. A man clambers up the cliffs, and sits in some crevice frequented by these birds, which are nocturnal in their habits. Just before dawn the guillemot returns from sea, and not seeing the man flies against him, and is instantly grasped and despatched. Others follow and meet with the same doom. Sometimes the man is kept so busy that he is obliged to squeeze one or two under his elbow until he has time to wring their necks. In a short time he is half hid in a heap of the slain. A boat waits below to carry away the spoil. The note of the lavy when he sits in his cave is like the squeaking of a pig when it sees the wind.

I did not learn much about the razor-bill (*chenalopex torda*), except that he burrows in the ground, and is caught by dogs. Both he and the guillemot are used for food.

There are no solan geese in St Kilda, but Boreray and the adjacent stacks are thronged with them. The shag (*graculus cristatus*) is pretty common in St Kilda, but is not used for food.

A solitary heron is often to be seen in the bay. Of land birds the starling is the most numerous. I have also seen a good many snipes. The village is haunted by hooded crows, and I observed a hawk more than once attacking them, and leaving one after another gasping and panting upon a stone, with gaping bill and drooping wings. Ravens, I heard, frequent the Dun, but I saw none.

On the 13th June I went in a boat that conveyed a party of men to the island of Soa to pluck sheep, and was surprised to see them ascend the cliffs by paths that one would have thought a cat would have had some hesitation in trying. But as I went to the top of Soa on my second visit, I shall defer giving an account of that island until afterwards.

On the 15th of July a boat took a number of young women to Soa, and left them there to remain for three weeks, to catch puffins for the sake of their feathers. On the following day another detachment (seven in number) were carried to the island of Boreray for the same purpose, and I went with them, and returned with the boat and men. On our way thither we went round by the cliffs on the north-east side of St Kilda, where the crew dropped their long lines into the sea.

In honour of the ladies I took the *piob-mhòr* with me, and, sitting in the stern sheets, played some airs, to which the men kept time with their oars. The day being calm and sunny, the crew soon grew heated with their work, when the girls took their bright coloured handkerchiefs and covered the heads of the men. They also arose from the bottom of the boat,

D

where they had been demurely squatting, and lent a hand in rowing, partly, I fancied, to give vent to their suppressed spirits.   A jolly and picturesque crew !

Boreray is 3¼ miles to the north-east of St Kilda. It rises to a height of 1072 feet above the sea, and is 1¼ miles in length.   It is faced on all sides by cliffs, some of which ascend to the top; but in some parts it soon grows less precipitous, and the sloping summit is covered with grass, which affords pasture for a flock of sheep belonging to the natives.   The west end rises into rocky heights.   Boreray is the great resort of solan geese, thousands of which were flying over our heads as we approached it.   We landed at a sloping cliff on the south-east.   One man, as usual, secured by a rope, leaped from the boat upon a little jutting rock, and another followed, and crawling up the cliff held the line firmly.   The girls in succession then jumped into the arms of the man at the foot, who lifted them on the slope, where, by the help of the rope, they attained a level spot. Some of the men accompanied them to help to carry the baggage, and I scrambled up with the others. By winding and hazardous ways we ascended to a height of perhaps 500 feet above the sea, where the ground was comparatively level and covered with turf.   Here there were a number of *clætyan.*   The

Old house in Boreray

girls were all laden with straw bags filled with stores, and one carried a kettle of burning turf in a basket on her back; but although they complained a little of being *blàth* (warm), they stepped up the perilous ascent like goats, without the slightest symptom of fatigue or fear. "Custom had made it to them a property of easiness."

Whilst we sat resting on the grass for a few minutes the dogs had already begun to catch puffins, sniffing and scratching their holes, and seizing the birds when they fluttered out. Whilst the sagacious animals pawed at one hole they kept a watchful eye on the burrows adjacent, as if they expected the puffins to issue from them. Some of the girls at the same time were plunging their hands deep into the holes and dragging out the birds, and twisting their necks with a dexterity which only long practice could give. In a short time all were bounding about with birds dangling from their waists. But although the dogs are useful auxiliaries, the women do not altogether depend upon them, but set snares—ropes with nooses attached—by which each will bag several hundreds a day. A few yards along from where we rested the puffins were sitting in countless numbers, and the air was all alive with them and solan geese. Although subjected every year to such wholesale slaughter, the sea-fowl are said to be increasing, probably in an inverse ratio with their human destroyers.

From the hill-side I could see, far across the waves, the Long Island, a dim diluted blue.

There are three houses on this island, which are occupied by the unprotected females, who come to "make feathers," as they call it. I did not see them at this time, but I inspected them on my second visit to St Kilda, and to make the picture complete, will insert the description here. These houses are all much the same, and externally look like little green knolls, but they are built of stone, on the same principle as the *clætyan*. One of them measures 15 feet long by 6 feet broad, and is 6 feet high at the hearth, which is close to the door. A semicircular stone seat is close to the hearth, and the space behind is a foot or so higher. This is occupied as a bed by the women. A hole in the roof above the fireplace serves as a vent. The door is about 2½ feet high, and has to be entered on hands and knees. These houses are said to be old.

Of all occupations for women, this of catching birds seemed to me the strangest. I noticed that some of the girls had their Gaelic Testaments with them, and was told they would have family worship as when at home.

In wishing me *Beannachd libh,* some of the young ladies expressed a fear that a yacht would call at St Kilda in their absence, and that I would go away in her, and they would see me no more; for which compliment I blew up the pipes when I got down to the boat, and played part of *Cumha mhic an Toisich,* and other laments.

Close to Boreray are two great detached rocks, which are literally white with solan geese. It has been calculated that these devour 214,000,000 herring per annum, which is equal to 305,714 barrels—more than the total average of herrings barreled at all the north-east stations. Yet no attempt has been made to exterminate the solan geese, and to secure the treasure which they consume! If herring fishing were started in St Kilda, a revolution (the whole consequences of which it is impossible to forsee) would occur immediately. That it would be a good thing for the proprietor, as well as for the ancient and interesting community that now inhabit the rock, one may safely predict.

The solan goose comes to Boreray and adjacent stacks in March, and leaves in October. The female lays one egg, and the young one (called *goug* in St Kilda) is of a dark brown colour until he is about a year old, when he turns white, excepting the tips of the wings, which are black. His bill is long and sharp, and slightly crooked at the point. He builds his nest with grass, sorrel, and anything he can pick up on the waves, such as shavings, sea-weed, rags, etc., and steals if he has a chance from his neighbours. About the middle of March the St Kildians launch their boat and go to Bóreray and Stack Armin to catch the old birds in the dark. Two men fastened at either end of a rope ascend the rocks, and on all fours crawl along the ledge where the geese are resting. The latter have always a sentinel posted, who, if he thinks all is well, cries "Gorrok! Gorrok!" on hearing which the fowlers advance; but if the sentinel cries "Beero!" the men remain motionless, with their bonnets drawn over their brows, and their faces on the rock. If the sentinel fancies it was a false alarm, and again cries "Gorrok!" the first fowler progresses until he is near enough to grasp the sentinel, and twist his powerful neck. The sentinel gone the whole flock fall into a state of panic and bewilderment, and crowd upon the man on all sides. He has nothing to do but despatch them. But it sometimes happens that the whole troop take flight with a "Beero! hurro! boo!" when the men have to crawl back without any game for that night. I received this account from Donald Og, who was a good cragsman in his day, although he is now old and crippled by an accident. In

May the St Kildians pay another visit to Boreray for eggs. The *goug* is ready for the table in September. Its flesh is said to be wild. The young geese are knocked on the head with a stick in the daytime.

CHAPTER IV.

The Government—Trade—Departure from the island.

ST Kilda by a legal fiction is in the county of Inverness, but it is virtually ignored by the British Government. The inhabitants, it is true, pay taxes on their tobacco, whisky, etc., and the Receiver of Wrecks at Stornoway claims the half of the flotsam and jetsam. But for this they might doubt if they are subjects of the Queen. They are too poor and too white-skinned to be deemed worthy of attention. The proprietor, MacLeod of MacLeod, who bought the island several years ago, for, I believe, £2000, may say like Louis XIV., "The State—I am the State," for his power is unlimited. The trade is a monopoly in his hands, and his serfs are obliged to deal with him on his own terms. It is true he has offered to allow them to go and trade where they choose; but he knows he might as well tell one who has been fettered until his limbs have lost all ability that he is at liberty to run, or bid the ostrich lift its wings and fly. When it was suggested by the *Spectator* that the inhabitants were entitled to a post, no one but MacLeod objected. At present there is no means of getting a letter sent to the island, except by favour of him or his factor, and pestilence or famine might be there, and the inhabitants have no opportunity of letting their condition be known to their fellow-countrymen. If it is right that any private individual should be allowed to possess and govern a remote island like St Kilda, surely it ought to be one who is thoroughly acquainted with the people, and who feels a deep interest in their welfare. But MacLeod has never had his foot on the island, and all his information is derived from hearsay. Even his agent cannot, except as regards business, have any knowledge of the subject, as he only goes there for three days per annum, and is busy all the time bartering and settling accounts.

Where one man has absolute power, it will be a rare case if he does not abuse it. This is proverbial, and is well exemplified in the dealings of the stewards or factors with the natives of St Kilda. Martin says that when the steward visited the island in his day, he was wont to bring a large body of hungry retainers with him, amounting to fifty or sixty persons, the leanest he could find in the parish, who were fed and fattened

for about three months at the charge of the poor
St Kildians, who, although fond of strangers, were
delighted to see their backs. What they could not
eat they carried off with them, leaving the natives
nothing to live on but sea-weed and coulternebs.
Previous to 1697 the oppression was still more severe.
At a much later date the islanders were permitted to
keep cows, but the steward claimed all the milk,
except at the season when the animals had none,
when the poor people were allowed to milk them if
they felt inclined. There has been a great reform in
this respect. The proprietor receives nothing now-a-
days but his rents, and the exorbitant profits on his
double-barreled trade monopoly.

Still I cannot help saying, that I should like to see
free trade established, and a regular and independent
postal communication opened up between St Kilda
and other parts of the kingdom. I was more than
once requested by all the men in a body to let their
position be known, in the hope that some one might
be so generous as to present them with a boat large
enough to carry a crew and cargo to the Lewis. It
has been suggested that the Gordian knot should be
cut,—that the whole population should be removed to
Canada. But I· think it is bad policy to transport
well-behaved people. Emigration is a panacea which
some prescribe for all evils. They remind one of the

novelists who always end with marriage. Neither gives a hint of what happens afterwards. It is a clumsy way of curing corns to cut off the feet. With free trade and a post, the St Kildians might be as happy at home as they would be in a colony. Besides, a large proportion. (say twenty-seven) are over forty, and too old to transplant.

From inquiry at every individual'(all save one gave me the information willingly) I have ascertained the amount of

Articles Exported from St Kilda in 1875.

		Prices given by Factor.
		s. d.
Cloth, .	. 227 yards of 47 inches and thumb, at	. 2 3
Blankets,	. 403 „ „ „	. 1 10
Fulmar oil,	. 906 pints (each pint equal to 5 pints Imperial),	1 0
Tallow,	. 17 stones 6 lbs. (the stone containing 24 lbs.),	6 6
Black feathers,	87 „ 15 „ „	6 0
Grey feathers,	69 „ 19 „ „	5 0
Cheese,	. 38 „ 6 „ „	6 0
Fish,	. 1080 "marketable," . . . each	0 7

It must be remembered that a large profit is charged by the factor on all imports.

I had seen as much of the island as I wanted, and had just begun to feel weary, especially of housekeeping, when, on the morning of the 19th of July, I was awakened by several voices bawling in at my bedroom window, "*Saothach! Saothach!*—A sail! a sail!" and springing up, and rushing to the door, I beheld a

yawl beating up against wind and tide at the entrance
of the bay.   This was the first sail I had seen near
the island since the factor's smack left six weeks
before.   A boat was launched and went off to her,
and all the village was astir.   Women could be seen
running from house to house discussing the great
event; a group of old fellows sat upon the ground
with their backs against the wall, and their bright
eyes fixed on the vessel.   On the shore stood Mr
Mackay, in his best rig, watching the motions of the
stranger through an opera-glass that made you see
double for the time, and gave you a squint for some
minutes afterwards.   I made my breakfast, and
packed up my bagpipes and other articles.   Presuming that I was about to leave them, all the women
flocked to my quarters, each with a small ewe-milk
cheese or a pair of stockings in her hand as a parting
gift; and in the midst of their "*Beannachd leats,*" a tall
gentleman, followed by the minister, stepped into the
crowded room.   He proved to be a baronet, the
owner of the yacht *Crusader*, who at once agreed to
give me a passage to the mainland, and whilst he
went up to the top of the cliffs to see a man go down
and catch a fulmar, I finished my preparations, and
was ready to go on board when he returned.   A snug
state-room was assigned to me, and from comparative
discomfort I was suddenly in the midst of luxury.

Anchor was weighed in the afternoon, and a light breeze wafted the little yacht out of the bay, but left her for several hours becalmed between St Kilda and Boreray. I had thus an opportunity of taking a last fond look of that precipitous island, where I had spent nearly seven weeks on the whole so pleasantly. The sun sank behind the wave in dazzling splendour, when the wind arose, and St Kilda soon vanished astern. I took the helm for several hours during the night, and steered by the compass and the moon, and congratulated myself on the lucky accident that had let me out of the trap. Next morning we were off Barra Head, 68 nautical miles from St Kilda. I found that the owner of the yacht was not only a good seaman, who lent an active hand in working the little craft, but a skilful navigator. The wind became fitful—now a momentary puff, and anon a dead calm. Keeping outside the Hebrides, and passing close to *Sgir-mhòr,* or Skerryvore, lighthouse, we approached Rathlin Island, off the north coast of Ireland. The last time I had been in this quarter was during a strong gale, when the waves were rolling like hills, and the barque I was on board of staggered under double-reefed top-sails, her lee bulwarks buried in foam, and the ruffian billows dashing in cataracts over the weather side. Now the land loomed softly through the still and sunny air, and the glassy sea, reflecting the clouds, might have

deluded one into believing that a passage to St Kilda
might have been accomplished with perfect safety
in a birch-bark canoe. The *Crusader* arrived at
Gourock, on the Clyde, on the 22nd, the whole distance sailed being 285 nautical miles. I was invited
to spend the night on board, and received every kindness during the passage. This gentleman sent two
clocks to the island, the first that had ever been there,
and other useful articles.

Next morning I proceeded to Glasgow, where I got
the first newspaper I had seen for two months. I
found it difficult to realise that I had never been out
of Scotland all that time, so different is everything in
St Kilda from other parts of the kingdom. I was
dressed in a suit of the native cloth, and could observe
by the noses of the ladies in the railway carriage that
I had brought with me the strong and peculiar odour
which adheres tenaciously to everything in Hirta, and
which arises from the turf smoke.

SECOND VISIT, 1876-77.

CHAPTER V.

I get a boat and take her out—Stornoway—Loch Erisort—
Scalpa — Obe—St Kilda revisited — Boreray — Infant
prodigy in Edinburgh—Subterranean dwelling—Discovery
of stone implements—Antiquities—The stone girl.

N my return from St Kilda I had no intention
of ever going there again, but I remembered
the promise I had made to the people to
try and get a boat for them. I knew that the
handsome craft they had received from Mr Young
was more suitable for the Clyde than for the strong
seas between St Kilda and the main, but believed
it was possible that a boat might be built that would
make a passage in safety. In the olden time the
natives thought nothing of rowing to Pabbay for
whisky or tobacco. Not having the means myself,
I made an appeal to others—wrote a letter to the
Scotsman, calling public attention to the isolated

condition of St Kilda, and the desire the people had for a boat that would take a crew and cargo to the market. As this had no result, I was tempted to try and raise the wind by subscription. By the kind assistance of my friends, and by the generosity of strangers, I succeeded in getting enough money to purchase a boat, and to defray the expense of getting her taken to St Kilda. Afraid that the builders on the Clyde would follow the Loch Fyne skiff model, and make a boat like Mr Young's, I employed a man at Ardrisaig who had been at St Kilda, and knew the seas that had to be encountered there. He was strongly recommended to me, and I gave him the order, and received a drawing that seemed satisfactory. The boat was built, and I determined to go out and see her safely delivered. So on the 30th of May 1876 I went to Lochgilphead to take possession of her. I cannot say that I was altogether pleased with her appearance—she was broad enough, but too lean in the quarter; but others praised her, and I tried to believe she was all right. I employed two men to take her through the Crinan Canal, and on the 1st of June I arrived in Oban. Here the men got drunk and extortionate, and I dismissed them. On the 2nd, my boat was lifted on board of the *Clansman* steamer, and carried to Stornoway. This was done by the Messrs Hutchinson free of charge. They had also instructed their agents and officers to give me every assistance on the route, for which I felt grateful. Although the steamer was full of ministers returning from the Assemblies, we arrived without any accident at Stornoway on the 3rd. Here I was introduced to Captain M'Donald of the fishery cruiser *Vigilant*, and was in hopes that he would take my boat and myself to St Kilda; but although he had no work to do, he was not permitted by the Board to leave his station. I should have liked to have seen Stornoway in a successful season, with all the boats and people in a state of activity; but there were no herring to be got, and about 400 or 500 large boats lay at anchor in the Loch. The streets were crowded with as many thousand fishermen, some of them listening to roaring preachers, and others walking listlessly with their hands in their pockets. A large number of women were also to be seen, sauntering idly, some of them in a state of destitution, in the streets. Fish-curers, who had advanced large sums to the fishermen to retain them on the spot, in the hope that the fishing would improve, looked the picture of despondence. A cloud hung over the place, metaphorically as well as literally. The wind blew from the west with heavy showers, and until it changed it was impossible to proceed to St Kilda. Whilst waiting for a fair wind, let me

E

narrate a mysterious occurrence, bearing on my narrative, that took place some years ago.

In the month of April 1874, a boat left St Kilda for Stornoway with a woman and seven men on board. Every man had a chest and the woman a small box, and they took provisions with them, and some salt-fish and home-spun cloth to pay expenses. The islanders went up the hill called Oswald or Osimhal, and watched the boat for several hours. All seemed well, but in the afternoon the weather grew stormy. The woman in the boat intended to visit some relations at Loch Inver.

On a Sunday, about a month afterwards, three London smacks entered the bay, and brought the news that the boat was lost near Lewis with all on board. Never doubting the truth of the intelligence, the inhabitants gave vent to their grief without restraint. The three skippers came on shore, and beguiled the time by playing quoits with flat stones, and when they witnessed the grief of the bereaved St Kildians they howled in mockery. I tell the tale as it was told to me; but I have recently been informed by a lady who was on the island at the time, that the three skippers behaved very decently, but that some of the men may have mocked the people. There was no minister in St Kilda at that time, but a catechist called

Kennedy filled the office. Although he understood English as well as Gaelic, he never thought of taking a note of the names of the smacks. The St Kildians say that the crews belonged to London, but that one man could speak Gaelic. Some time afterwards some of the clothes of the missing men were brought to St Kilda by the then factor, and were said to have been found in a cave at Lewis. The money that had been sewed into the pocket of a certain waistcoat was amissing. The people got gradually resigned to their fate, although I heard them on my first visit declare that they believed the lost crew had been murdered. But I thought at the time that this was a preposterous suspicion, which could only be entertained by people living in solitude, and ignorant of the world outside. But strange to relate, I was told by Mr MacIver, banker in Stornoway, that a letter had been received from a firm in the Transvaal Republic, by the minister of Harris, stating that Donald MacKinnon, one of the lost crew, had just died at Pilgrim's Rest, Lydenburg Gold Fields, of a fever, and had left property to the amount of £37. On my expressing a suspicion that the strange story might be untrue, Mr MacIver informed me that the money had actually been lodged with him.

Why Donald MacKinnon had never written to

inform his father and other relatives of his fate, is a mystery that none can fathom. But if he was preserved, it is possible that some others of the missing crew may have been also saved. I may mention that Sir John MacLeod, then proprietor of St Kilda, caused an inquiry to be made at Lewis at the time, but without eliciting any information.*

On the 12th I made a trip on board of the *Vigilant* to Loch Erisort, a few miles to the south of Stornoway. We anchored near the mouth of the Loch, and Captain M'Donald took me in his gig to see a churchyard situated on an island called St Colme. It is quite close to the water, and is about sixty feet square. It might have been originally large enough for the district, but now the accommodation is shockingly insufficient. Although there seems plenty of suitable ground outside, the people persist in interring the dead within the ancient limits. Nay, not interring, but piling the coffins one on the top of the other, until they have risen to a height of ten feet above the surface. The coffins are not even covered with earth, but are only wrapped in turf.

* The relatives of a Donald MacKinnon, a native of Barvas in the Lewis, who was known to be in Africa, are claiming the money, which has been paid to the father of the St Kilda Donald MacKinnon. Alas! for the poor St Kildians, their hopes have been excited in vain.

In some places they look like the steps of a stair covered with a carpet. One can count the tiers. As Captain M'Donald stepped before me, I expected to see his nautical, square-built figure sink, like the ghost in Hamlet, through the hollow turf. I myself felt as if I were walking on thin ice, which might give way any moment and bury me in corruption. I was surprised that no foul smell pervaded this charnel pit, until the Captain pointed out that there were two holes made in every coffin by the rats, and that a body was no sooner left than it was devoured. He poked his cane, like a tide-waiter, into a new coffin, and found nothing but bones. The place swarmed with rats. We could hear them in the still evening air squeaking and fighting over their horrible banquet under our feet, and a shoal was running along the beach to whet their appetites with a limpet and a breath of fresh air. Some of the sailors, full of morbid curiosity, knelt on the coffins, and with their heads upside down peered into the holes. There is a small chapel in ruins in the midst of the ground, but its walls are half buried in graves.

On the 13th I went on board of the *Vigilant* to the island of Scalpa, where I wanted to be left for a change. Here I was hospitably entertained by Mr Campbell, the chief man on the island. His house (called the *Taigh Geal*) is built on the site of

one in which Prince Charlie found shelter in his wan-
derings.   An inscription in Gaelic cut on the lintel
records the fact.   The population of Scalpa is about
four hundred, and the people subsist chiefly by
fishing.   The country is hilly, and the land consists of
rock, stones, and bog ; but wherever it is fitted for
cultivation, it is made to bear a crop of potatoes.   The
Highlanders are often taunted with indolence ; but
those that call them so forget the difficulties they have
to contend with.   In this island I found them polite
and hospitable.   In one of the huts I saw about a
dozen girls thickening or felting blankets.   This
they do by tossing the cloth about upon a board,
and sprinkling it with dirty water.   One of the girls
sang a Gaelic song with great energy, and the
others joined with all their hearts in the chorus.
On my entering the room the songstress on the
instant composed or altered a verse to suit the
occasion.   I was interested in seeing this ancient
Highland custom.   In St Kilda the women do not
sing, but hiss at this work.

On the 17th the *Vigilant*, which had returned to
Stornoway, again called at Scalpa, and took me
and my boat on to Obe, in the Sound of Harris.
The navigation of these straits is considered difficult,
but Captain M'Donald made light of it, and as if
to display his seamanship, ran down the Sound, and

then tacked up again, passing within a few feet of sunken rocks, which were indicated by water of an ominous smoothness. At Obe I found my old friend the *Janet* lying weather-bound. Here also I saw two St Kildian women, who had come to Harris nine months previously to be confined, and were yearning beyond expression to be home again, never having heard from their husbands since they left. They looked thin, and grumbled afterwards at the Harris fare, which consisted chiefly of tea and bread. No fulmars, no solan geese, no mutton. Ah! *àite bochd.* On the 18th the *Vigilant* left for Stornoway.

I found quarters in the humble inn, where civil treatment and plain but good fare, with moderate charges, made up for indifferent accommodation. In one of the huts I heard a man "discourse most excellent music" on the chanter to a group of appreciative listeners, one of whom whispered to me, "He is the best piper in all Scotland." Pipers are more fortunate than prophets, and are often most honoured in their own countries.

At Obe I engaged two men to work the boat to St Kilda, agreeing to pay them L.8 for the trip, but stipulating that when the boat reached the bay they were to have no further claim upon me. This seems a large sum for such a short passage; but for all I knew they might have been detained on the island for weeks. The skipper of the smack who was present, and had remained silent until the bargain was closed, then agreed to give them a return passage.

On the 22nd the wind changed to the east, and about seven in the morning we set sail, two men and myself in the new boat, and a rope connecting us with the smack. I was opposed to this arrangement, and wanted to make the passage in a separate and independent way; but the men and the skipper were old friends, and wanted to go together for sociality. In calm weather the boat would have beat the smack, and if the wind had increased we were to be thrown off; so to be taken in tow was no great favour, although MacLeod cast it in my teeth afterwards.

When about half way St Kilda was descried on the western horizon—"suspected more than seen"—for although the day was bright and sunny, a thick haze obscured the distance. We reached the island about five in the evening, and separating from the smack, cast our anchor near the shore. Soon a crowd began to gather on the rocks, but they were in no hurry to launch their boat. I observed one of the women who had come with the smack standing on deck, and holding up her infant (born during her absence) in a triumphant manner, although she was too distant to be seen from the shore. At length a boat is pushed off and pulls towards us; the crew stare doubtfully at me, and then, as they come alongside, repeat my name, and grasp my hand with a hearty "*Cia mar tha sibh—'S math mar a tha sin!*" I and the two Harris men jump into the shore boat, and are landed on the rocky bank amidst a crowd of men and women. But whilst I am busy shaking hands all round, one of the Harris men discloses the story about the lost boat and Donald MacKinnon, and in a moment all is confusion, grief, and amazement. Women sit upon the ground and chant their lamentations, and men stand with open mouths and eyes, and mutter, "*Sgeul ionngantach! Ah! cruaidh!*" Strange news! ah! cruel!

But the boat goes off to the smack and brings the two women on shore, where they are received with conjugal kisses. Ten months had elapsed from the receipt in Harris of the letter from Africa until it reached St Kilda, although the one place can be seen from the other in a clear day.

I took lodgings in one of the cottages for a week or two; but afterwards got a house again to myself, and lived all alone. On the 24th the *Janet* left the island, and for the next eight months no vessel ever came near it. Before leaving the factor presented a document to the men who were gathered on the shore, and requested them to sign it if they wanted MacLeod to send supplies in Autumn as usual. He also stated that he would allow the St Kildians to take a cargo in the new boat to Harris, and sell it, and if they got more for the produce than he gave, he would willingly raise his prices. The men seemed determined to try and trade for themselves; but the minister (who the previous year had been grumbling at the prices charged for supplies as much as any one, and wishing the islanders had a boat large enough to go to the market in) interfered, and persuaded the men to sign the paper. I was not pleased at this, as the boat, although she was found useful in making excursions to Boreray and Soa, had been got on purpose that the islanders might have an opportunity of trading with Harris. I saw that all my labour—and it had been great—in collecting subscriptions, and all the money I had collected from benevolent people, had, through the uncalled for interference of this reverend gentleman, been thrown away. The St Kildians having now no need to go to Harris themselves, never thought of giving me a passage back, but trusted to a yacht calling. I had some oatmeal and tea and sugar. I bought cheese and milk, and the people gave me a present of a sheep, so that for three months I had as much food as I desired. I spent the time in wandering about the island, in making excursions with the men in the boats, in playing the pipes, in sketching, taking notes, and trying to learn Gaelic. On Sundays I sometimes went to the church, and sometimes remained in the house, pacing from one corner of the room to the other. Very often I did not hear a word of English or Gaelic from Saturday night to Monday morning.

On the 29th June I went with a party in the new boat to Boreray, and the men were all delighted at her speed, and at her going so close to the wind. All the men but two went up the cliffs, and I was tempted to go along with them. Tied to the end of a rope fastened to a man who preceded me, I clambered up such paths as one may see in a nightmare. The way I had gone the previous year seemed a turnpike in comparison. Sometimes I was indebted to my guide for a pull up some difficult bit ; [and I succeeded in reaching the top. A house of the bee-hive pattern, and of a very superior order, formerly stood on this island ; but I was grieved to find that it had been utterly demolished, and used as a quarry to build *clætyan* with. It was inhabited by a hermit. Martin says : " In the west end of this isle is Stallir House, which is much larger than that of the Female Warrior in St Kilda, but of the same model in all respects. It is all green without, like a little hill. The inhabitants there have a tradition that it was built by one Stallir, who was a devout hermit of St Kilda; and had he travelled the whole universe he could scarcely have found a more solitary place for a monastick life." Old men remember it well, and say they have often slept in it. There is no trace of the Druidical circle that is said to have existed in Boreray. The men who had come with me dispersed to catch fulmars, and I was left in charge of a youth called *Callum Beag;* but I grew tired sitting on the top of the cliff, and in spite of my guide's remonstrances, and the shouts of the men in the boat, I ventured to descend, and fortunately reached the boat in safety. One gets familiar with great heights sooner than would be credited.

One day shortly after my arrival an old man, happening to be up the hill at the back of the village, descried, what he imagined to be, two marks cut on the turf on the top of Boreray. A party of men had gone to that island about a fortnight before to pluck sheep. He came down to the village in great distress, and communicated the intelligence to the rest of the people, who, to my surprise, were thrown into a state of consternation. The women sat upon

the ground and began to lament. On inquiring the
reason I was informed that a system of telegraphy
had been long established in St Kilda, and that
two marks cut in the turf in Boreray signified that
one or more of the party was sick or dead. I
went up the hill, and with a glass discovered that
one of the marks was a number of men building
a *clætya*. I explained this to some of the people
who had followed me, but failed to convince them
for a time. In the evening, however, when the
boat returned from Boreray with the plucking party
all well, the sceptics acknowledged with joyful smiles
that my glass was better than their eyes.

About this time the minister asked me if it was
true that a child had been born in Edinburgh, which
spoke very good English immediately after birth,
and foretold that this year was to be a bad year
for Hirta, and that after uttering this prophecy it
expired. A man in Harris (the wag!) had read it
out of a newspaper to one of the St Kilda women.
"Tut! I don't believe it myself," said the minister,
"but the people wanted me to ask you if you had
heard of it." Whether that child was ever born or
not it certainly spoke the truth. The inhabitants
of Harris and those of St Kilda have no great
respect for each other. The Harris people call the
St Kildians *gougan* (young solan geese), and mothers,

when their children are naughty, threaten to send them to *Hirta*. The St Kildians, again, never mention Harris but in derogatory terms. *Tha e àite bochd—mosach—salach.*

There were no fish to be had. The crop, from the cold wet weather, was all but a failure, and MacLeod sent no supplies,—a bad year for Hirta, indeed; but it is always darkest before the dawn.

On the 13th of July I determined to examine a subterranean house that was said to exist at the back of the village. The door was covered by a crop of potatoes, and the owner of these was very reluctant to have them disturbed; but after a deal of persuasion, and on my offering to pay the damage, he at length acceded to my request. This underground house was discovered about thirty-two years ago by a man who was digging the ground, and after a cursory glance was covered up again, and never opened since. It is called the *Taigh an t-sithiche*, or House of the Fairies.

After a little search the entrance in the roof was found. Two men volunteered to clean out the stones and earth that had fallen into it, and they worked with a will. It was ascertained to be 25 feet long by about 3 feet 8 inches wide, and about 4 feet high. The stones with which it is built are large, and the massive walls converge towards the top, so that lintels can be placed on them to form the roof. At a right angle with the house is what I thought to be a passage, but the men declared it to be a *crupa* or bed-place. The floor was covered with peat ashes to the depth of a foot or more, and embedded in these I found the bones of cattle, sheep, fulmars, solan geese, and puffins, together with the shells of limpets. These showed that the occupants of this primitive dwelling had lived upon the same sort of diet as the islanders at the present day. The presence of solan geese bones proves that the people must have had a boat of some sort, as there are none of these birds in St Kilda. I also found numerous fragments of rude pottery blackened with soot on the outside. No pottery has been made in St Kilda within the memory of man, nor is there any traditions concerning it; but the men told me they had often found vessels of coarse earthenware when digging on the site of old buildings. I also discovered the fragments of a stone lamp, and numerous stone knives and cleavers, as well as some stones that were probably used for pounding corn. I found some stone axes stuck around the aperture in the roof, from which I concluded that this had been either the door or the chimney; at all events, that it had been open when the house was occupied. This entrance commands a view of the bay. When digging near this sub-

terranean dwelling some time afterwards I chanced
to find the kitchen-midden of the establishment,
which contained a large quantity of the same refuse
as I found on the floor, including a stone implement
or two, and some fragments of pottery. I afterwards
discovered numerous stone implements on the surface
of the ground in the ruins of old houses, and feel
gratified to think that I have been the first to do so,
and have been thereby enabled to throw a little
original light upon the history of this interesting
island. The *Taigh an t-sithiche* is probably of great
antiquity, although it is impossible to conjecture the
date. I have no doubt that stone implements were
used at a very recent period in St Kilda, probably
when knives of good steel were common in Harris
and Skye. Some men still alive say they have seen
an ox knocked down with a long stone, and others
have used stone lamps when in Boreray. I myself
have seen a huge stone taken to break drift-wood. I
forgot to mention that there was a drain under the
floor of the subterranean house.

In Glen Mòr there is a curious old building, which
was almost entire on my first visit, but has been much
damaged since by two men who wanted the stones
because they were *briagh* (pretty) to build *clætyan.*
It is now called *Airidh mhòr.* Internally it is circular
in form, and about 9 feet in diameter, and at the top

the stones overlap and form a dome. It contains three beds in the wall. It was occupied, according to Martin, by an Amazon, who used to hunt, before St Kilda was an island, all the way to Harris. Outside it resembles a little green hill.

Near to *Airidh mhòr* is a spring of excellent water, called *Tobar na buaidh,* or Well of Virtues—miraculous virtues. An altar formerly stood near it, on which people who came to drink placed offerings.

At the back of the village is a stone, which does not differ in external appearance from the numerous stones scattered around, but which was supposed to possess magical properties. It is called *Clach an eòlas,* or Stone of Knowledge. If any one stood on it on the first day of the quarter, he became endowed with the second sight,—could "look into the seeds of Time," and foretell all that was to happen during the rest of the quarter. Such an institution must have been of great value in *Hirta,* where news are so scanty. To test its powers I stood on it on the first day of Spring (old style) in the present year, but must acknowledge that I saw nothing, except two or three women laden with peats, who were smiling at my credulity.

There is an old cellar, or *clætya,* said to have been built by one man in a single day. Some of the stones are too ponderous to be lifted by any two men of

F

those degenerate times. The people refer to this cellar as a proof of the superior strength of their ancestors.

There were formerly three chapels in St Kilda, dedicated respectively to Christ, Columba, and Brendon. Not a vestige of them is now to be seen ; but old men remember when the ruins of one, 16 feet in height, stood in the churchyard. One of the stones, having a cross incised on it, may yet be seen built into one of the cottages.

On the side of the hill that overlooks the bay, and is called *Sgal* or *Sgar*, and amongst the great mass of stony chips that have fallen from its rocky ribs, there is a subterranean house, which had been covered up and re-discovered this year. I threw out the rubbish, and found that it contained two beds in the wall ; but I could see nothing to indicate that it had been used as a permanent dwelling. Probably it was built for a hiding-place in times of danger.

There is a conical hill, like an extinct volcano, at the west side of the bay, called *Ruaidh Bhail.* On the summit a great number of huge blocks of trap have been piled up as if by some playful Titan for a prop. One of the blocks rests like a lintel on two others, and thereby hangs a tale. It is related that whilst a girl was engaged snaring puffins on the

Ruaidh Bhail.

cliffs about a mile from this spot, a strange man suddenly jumped to her from the island of Soa, another mile or so distant, and the prints of his heels are still to be seen in the place where he alighted. The girl (poor thing!) was so frightened that she leapt to this mount; but fell petrified on the awkward place in which she now lies. If she had been turned into salt she would have been used long ago, as MacLeod charges L.3, 10s. a ton for it, although it is to be bought in Scalpa for L.1, 4s.

CHAPTER VI.

Coulternebs—Lockjaw—The Dun, antiquities—Catching
fulmars—Soa, antiquities.

ON the 19th of July two boats took twelve
unmarried women to Boreray, to be left
there to "make feathers," for three weeks.
The blind man went in one of the boats to assist in
keeping her off the rocks whilst the crew went on
shore to catch birds. One of the boats returned in
the evening of the 20th with a large cargo of puffins,
which, contrary to expectation, were abundant in
Boreray last year. The blind man had sat for
thirty hours in the boat—a dismal time, but he
seemed quite cheerful. This might serve as a lesson
to many who grumble at trifles. "Take physic
pomp." On the 22nd all the people were busy
plucking birds, and the smell of roasted puffins—"a
very ancient and fish-like smell," came from every
door. Some of the men had caught five hundred
and twenty birds each.

On the 30th a fine infant died of lockjaw. The diet of the mother had been chiefly oatmeal. She had not tasted sea-fowl for some time. The infant, she confessed, had been fed with milk, water, and sugar, which seems to me unnatural nourishment and the probable cause of death. Whilst I sat with the poor convulsed child on my knee, and a crowd of women were squatting on the floor awaiting its death, the minister stepped into the room and said, " There is no hope for this infant—let us baptise it," and I shuddered to think what had become of the thousands of poor children who had died before a minister lived on the island.

I happened to see another infant that seemed to have the premonitory symptoms of *tetanus.* The relations were alarmed, but this did not deter them from trying to give the little sufferer another large draught of sugar, water, and milk. I persuaded the woman who had the mixture ready in a castor oil bottle without the neck, and was about to pour it down the infant's throat, to throw it out, and to give the child no more of it. The infant recovered, and is past the critical period. Something equally unnatural and deleterious may have been given before sugar was introduced, although the disease seems to be more virulent than it was. I mention this for the consideration of medical men who have manifested a laudable interest in this important subject. The population numbers seventy-five (1876).

On the 31st I went to the island called the Dun, which is only separated from St Kilda by a narrow channel, in a boat, with a crew of three men and three boys, who, for want of better work, tried to catch puffins. These, although the sea was dotted with thousands, were uncommonly shy and difficult to catch. Only about forty-five a piece were bagged. A pagan altar formerly stood on this island, but it has been demolished, and little but the site is visible.

At the southern extremity of the island is a mount, on which great blocks of stone have been piled up in wild disorder by the hand of Nature. Near to this, and the road is rough, is a small cave which the people call a *Sean Taigh*, or old house. The roof is so low that one cannot stand upright in it, but a doorway has been built at the entrance. This house, not made with hands, was probably used as a hiding place. It is still occupied by the people who visit the Dun to pluck sheep or catch birds. A steep crag arises at this spot and forms an impassible barrier to the extremity of the island, except to a St Kildian, whom no crag could stop. But this crag does not go quite across the isthmus, and a wall had been built to obstruct the passage. The masonry is of the coarsest kind, and the stones have been

Straits between St. Hilda and the Dun.

6 miles from Dun.

obtained on the spot. No mortar has been used. Some old men say they remember when there were loopholes for arrows in the top. The walls being thicker at the base leave platforms on which the defenders could have stood. A small cave, open at both ends, passes underneath the wall, and lean men might have been able to crawl through it. I examined this rude fortification with much interest, as it is the only existing building erected for purposes of war in St Kilda. The people had probably trusted to hiding themselves in case of invasion from pirates, who, according to tradition, were wont to visit the island. I believe I am the only stranger who has visited this spot (for Wilson only saw the fort through a spyglass), and the description of a castle given by Macaulay was founded on rumour, and elaborated from his own imagination. He says "the stones of which this strange fabric was constructed are large, nearly square, and must of consequence have been wrought out of a quarry, there being none of the same colour or substance to be found in the island above ground. It is plain that those who laid them together understood the rules of masonry much better than the St Kildians of this age, and they must undoubtedly have been men of greater power."

There is no such castle, and the natives, who must have heard something of it if it had existed so lately as 1759, are positive that there never was any other fort on the Dun than the rude structure I have described. This island, although of no great height, and although comparatively tame on the side facing the bay, is wild and picturesque where it faces the ocean. Some of the crags are crowned by pinnacles and fantastic protuberances, and the base is perforated with caves, into which the foaming billows rush and rage for ever. I observed large numbers of the laughing gull (*larus ridibundus*) sitting on the shelves of the rocks.

On my return from the Dun I found a boat at the shore that had come from Boreray with a cargo of puffins to be plucked at home, so as to assist the women, who were suffering from *Cnatan na Heric*, or Harris cold,—sometimes called the boat cold,— combined with swollen throats, and no wonder, for the weather had been bad, and these females had never changed their clothes, but had slept in the garments that they wore during the day; and although accustomed to severe exercise in the open air, had sat exposed to the cold, plucking feathers from morning to night. Their fingers had become so sore at this occupation, that they were obliged to pull out the tail and pinion feathers with their teeth. They suffer great hardships, and only get the pittance of six shillings a St Kilda stone (twenty-four pounds)

Boreray from the Dun.

for the feathers, which are of excellent quality. On
the 3rd of August a boat went to Boreray and
brought back a cargo of puffins and *gougan* or young
solan geese. On the 6th two boats went again to
that island and brought back the twelve young
women who had been catching puffins, together with
the feathers. I calculate that eighty-nine thousand
six hundred puffins must have been killed by both
sexes in 1876.

On the 11th Parliament assembled, and it was
debated whether it would be advisable to catch the
young fulmars, or to delay a day or two in the hope
that the weather would improve. It was decided to
delay, and in the meantime to bring out and test the
ropes used for going down the cliffs. Some of the
ropes were made from hair cut from St Kilda horses,
and were forty years old. Ropes of manilla and
hemp are now used, and fewer accidents occur than
in the olden time, when ropes of hair and even straw
were employed. Some of the men made me feel the
bumps and scars upon their scalps, caused by the
falling of stones from the cliffs above, whilst they
were dangling below.

On the 12th fulmar catching began in earnest.
I went in the morning of the 14th with a party of
men in a small boat to the island of Soa, which is
close to St Kilda. It is difficult to land on Soa in any

weather, from the swell of the sea and the steepness
of the rocky shore, but I determined to go to the
top. We landed on the south side. With the end
of a rope around my waist I leapt on the rock and
climbed up the cliffs at the base, a man preceding
me, also tied to the rope as usual. I got a pull from
my guide when needful. At a short distance up
the rocks became less regular. Great masses of
stone spring tower-like out of the ground, and blocks
of all sizes are crowded together on the steep
acclivity. The young man who had guided me up
the most dangerous part of the road, left me in
charge of an old one called *MacRuaridh*, or the son
of Rory, who, although he totters on level ground,
goes up the steep hill without any difficulty. About
half way up, where the ground is covered with great
stones, he showed me an antique house, which, tradition says, was built by a murderer called Dugan,
who was brought to Soa to starve to death. He
took shelter under a huge stone that springs out of
the ground at an angle like the chisel of a plane.
He then built sides and a front, leaving a door
about two feet square. It is related that Dugan's
bones and dirk were found some time afterwards in
his bothy. Numbers of puffins were sitting on every
stone around us, and I thought how dismal their
melancholy note must have sounded in the ears

of the dying *Dugan*, and how they must have re-
sembled fiends mocking his agony. His house is
still occupied by the women who visit the island
to "make feathers." One told me she had lived
six weeks in it, the weather being too rough to
allow a boat to come to take her and her companions
away. There are two or three other houses equally
rude and primitive in the vicinity, but they are
without associations. A little farther up the way
is barred by some land cliffs, but going round these
we reach the summit, which has a gentle slope, and
is covered with light green moss and dark green
grass tipped with vermilion. Here stands an altar,
in almost as perfect a state as when the St Kildians,
clad in sheep-skins, worshipped their pagan gods.
The holy ground seems at one time to have been
marked off by a circle of small stones, some of which
remain. The altar is built of loose stones, and is
about 3 feet in height, and square in form. It stands
in a spot well calculated to excite the religious senti-
ment, being some 900 feet nearer heaven than the
sea, and commanding a view of the cliffs of St Kilda,
with the ocean foaming inaudibly at their base. The
ground ascends with a gentle slope until it terminates
in a precipice 1031 feet in height. Far down I
could just distinguish two of our crew who were
busy catching fulmars on the rocks, and the boat

floating like a tiny mussel-shell at the base. These afforded a kind of standard by which to estimate the height of this stupendous crag. MacRuaridh and I sat and rested for a little on the edge of the cliff, but he soon grew tired of doing nothing, and began to peer down in search of young fulmars, some of which he saw on a cliff adjacent—a cliff broken here and there with grassy ledges. He asked me to hold the rope, and over he went, to my great anxiety, and brought up sixteen birds. We then returned by the same road to the rocks where we had landed. The sea had risen very much since then. After waiting for two hours the boat came round the island heavily laden with fulmars. Some of the crew had embarked at the other side of the island. But four or five came down the rocks to where we were, and cast anxious looks at the waves that came sweeping along from the west at a right angle with the shore. Two men sat on the top of the cliff, each holding a rope, by the help of which the others slid into the boat. Then came my turn. A line was fastened around my waist, and a hair rope put into my hand. I was peremptorily requested to take off my shoes; and as I descended I pushed my toes into any crevice or crany that offered, until the rock became so smooth that I could find no hold for my feet. Then I was obliged to be passive, and allowed myself to be lowered like a sack until I reached a small limpet-covered shelf on which the waves rose about knee-deep. "Jump! jump!" shout the crew; and when the boat mounts on the wave I leap and fall in a heap amongst the fulmars—all right. The air was quite calm, but the sea continued to rise, and the boat was in imminent danger of being dashed to pieces against the wall. At one time she became altogether unmanageable, and received several hard bumps. The water, too (she being heavily laden with fulmars and men), began to pour over the gunnel, and I thought every wave would send us to the bottom. It being impossible to get the two men on board at that spot, the boat was rowed along to a cliff farther south. The waves were quite as wild there; but a double line having been passed around a projecting stone and the ends held firmly in the boat, the two men slid down and pulled the rope after them. A few strokes of the oar took us out of danger—*mhuir fhiadhaich*—*droch chlata*—a wild sea —bad shore—muttered the men. A morsel of cheese and a bit of oat-cake was all I had tasted during the day, as I had hurried off before breakfast. It was dark when we arrived at the village.

CHAPTER VII.

Fulmar catching—Misery of the people for want of a mail—
MacLeod breaks his promise, and leaves the people to starve
—*Goill air a gleann*, false alarm—Spinning and Weaving
—Fairy tales—Lady Grange.

N the 16th of August I ascended the hill
called Connagher, where all the men had
gone to catch, and the women to carry
home, fulmars. The weather was very warm, and
although I carried my coat over my arm, I was fain to
stop on my way up, and cool myself in the light sea-
breeze. About half way up I saw my friend Tormad,
with his ruddy face and large white beard, seated on
the edge of the cliff, with his attention fixed on the
rope he held in his hand. "Who's below?" I asked,
as I sat down beside him. "Neil," he answered.
"Is he far down?" "Far," he replied. Neil's voice
could be heard calling from the abyss, "*Cum agad e*,"
"*Leig leis*," "*Thoir thugad e*,"—"Hold on," "Let go,"
"Pull away." In a little a crash and a rattle sound
from below, and Tormad looks anxious, and with

craning head listens intently; whilst two girls who had joined us step with their bare feet to the very verge of the precipice, and stare below.   One of them, who has a light, lithe figure, looks very picturesque, as she stands poised on that awful cliff.   She has a Turkey-red handkerchief on her head, and wears a coarse blue gown of a quaint pattern girdled at the waist, and only reaching to her knees.   Her limbs are muscular, and browned by the weather.   She is engaged to Neil, and naturally feels anxious on his account.   A shower of large stones had fallen, any one of which, had it chanced to hit him, would have knocked his brains out; but fortunately a projecting crag above his head saves him.   Tormad shifts his position to where he thinks the rock is less frangible. I leave him, and climb to where the rocks form a lofty head or promontory, which commands a view of the face of Connagher.   This hill rises 1220 feet above the sea, and is a precipice almost to the top. The foot had been cleared of fulmars on the previous day by men who had ascended from boats; now the work had to be done from above.

It is a dreadful trade.   A sound like the crack of a musket is occasionally heard, and one sees a huge stone bound and rattle with long leaps into the sea below.   Parties of two or three men laden with birds on their shoulders are seen climbing, by steep and perilous paths that would frighten a sleep-walker, to the summit. From the spot where I lay basking in the sun a path leads downwards to a grassy bank that slopes steeply to the edge of a cliff. This is considered a safe road for women, and a number of them go by it to where the men can bring them fulmars. Some of the girls can carry about 200 lbs., and seem proud of their strength; but as they toil up the dangerous path, I hear them breathing heavily and in apparent distress, but in a few minutes they are all right again.

In the intervals of business à number of them sit around me, and offer me a share of their oatcakes, cheese, and milk. A number of men also come up the path, with coils of ropes and bundles of solan geese stomachs on their backs. They are all barefooted and stripped to their underclothing. A pile of fulmars has been collected beside us, and the men, whilst they rest, economise time by extracting the oil. The receptacle for holding the oil is the stomach of the solan goose, which is held open by one man, while another takes a fulmar, and, squeezing the body, forces the oil in a stream from its gaping bill. When the fulmars and oil are carried home they are divided. The people sit up during the night and pluck the birds. The fulmar seasons lasts between two and three weeks, and all the crags of St Kilda as well as

adjacent islands are searched for them. I ought to mention that it is only the young fulmars that are caught at this time. As they are unable to fly, no art is required to catch them. At this dangerous and laborious work a man, with a female assistant or two, will during the season earn L.5 sterling at the most. He gets the flesh in addition. But L.5 in St Kilda is not equal by any means to the same sum in Glasgow. The upper part of Connagher is broken into vertical ridges and furrows of sublime magnitude. Some of the ribs descend like buttresses to the sea.

In the beginning of September I began to be slightly alarmed that no yacht would call, and that I might be detained a prisoner on the island for a year. My stock of provisions was exhausted, and I had to give up tea and coffee. No one but the minister had a supply of these luxuries. Shortly afterwards I had also to do without bread, and tasted none of any sort for five months. The meal the people had bought from the factor was all consumed about this time, or a little afterwards, and they had to depend upon their own crop, which, from the wetness and coldness of the summer, was all but a total failure. Some began to dread that MacLeod would break his promise to send supplies. I flattered myself that if his vessel arrived, I might get a letter sent home by her to let my friends know I was alive

G

and well. The people were not only anxious about
the arrival of food, but burning to receive further
intelligence about Donald MacKinnon. If he was
saved in some mysterious way, why might not some of
the rest of the crew have been preserved also? "Is
my poor wife alive?" "Is my mother, my brother, my
son, my father living or dead?" "Is my husband alive,
and is he married again? are all the women black
in Africa?" Such were the questions that passed
through the minds of the people, and often found
expression. Prayers were offered up in the church
for the preservation and safe return of the crew of
the lost boat, which were listened to with shrieks and
tears. Every time I went up the hill with my glass
I would be questioned by some one on my return,
and my answer, "*Cha 'n fhaca mi saothach*—I did not
see a ship," would be shouted from one end of the
village to the other. The weather in September was
very good, but MacLeod's vessel never appeared,
notwithstanding the solemn promise of his factor.
"*Tha e side briagh*," said every one, "*Thig saothach
an dheugh*—Beautiful weather; a ship will come
to-day;" but she never came. Some thought she
was waiting for the men who had been carried off to
Africa. My shoes, from clambering on the rocks and
stony hills, were in a shocking state; but afraid that
they might go to pieces altogether, and leave me

barefoot, I laid them aside, and buying leather from the *maor* or ground officer, I got one of the men to make me a pair of *brogan tiondadh*, which, when well stuffed with wool and sheep-skins, answered pretty well, although they would have excited astonishment in Princes Street.

The people plucked up their oats and barley, and thrashed them out. The grain looked like chaff, and when ground by handmills was like dust. It was, however, precious. The people brought me supplies by rotation, and would not accept money for it. Some said, "Send me something from Dun-eidionn." I took one handful for breakfast, and another for supper, which I made myself into gruel. I bought milk. I had a longing sometimes for two handfuls, but was obliged to abstain. I boiled a small bit of salt mutton and six small potatoes at 4 P.M., but felt a craving for more farinaceous food. I had an iron lamp, and sometimes bought and sometimes got presents of fulmar oil for light.

The grain when thrashed out is put into a tub made of straw, like an inverted bee-hive, and hot stones are thrown in to scorch it. It is then ready to grind with handmills. This work is done by the women, who, sitting on the ground, work like furies. A sheep-skin stretched upon a hoop, and perforated with a hot wire, serves as a sieve to separate the meal from the husk. During the winter evenings, two women grinding at the mill were to be seen in every house.

On the 5th of October, in the evening, whilst I was sitting alone in a cloud of peat smoke, gazing at vacancy by the dull light of an iron lamp, my door was suddenly thrown open, and a woman in a state of alarm bawled out that there were strangers in the glen. "*Tha goill air a gleann !*—Come, come up the town!" I went up with her, and found Parliament assembled, and all the people in a state of intense excitement. Five girls, it seems, had been in the glen, looking after some stray cows; in the dusk, when they heard a loud whistle (*feadun cruaidh*) several times repeated. They rushed over the hill, and reached the village in a state of terror. "Had it not been the whistle of a bird," I asked. "Oh, no! it was not like that—*Bha goill air a gleann.*" I then suggested it must have been shipwrecked sailors, if any one, whom it would not be right to leave in the glen all night, cold, wet, and hungry. This moved the women, and it was arranged that five men, armed with long sticks, should go up the hill and shout, and if there was any reply, return for me to speak English to the *goill.* Whilst this party marched up the hill, a number of us sought shelter from the rain in one of the cottages; and an old man related that he remem-

bered when a ship was wrecked 'on the island, and a
number of the sailors succeeded in landing in the
glen, and remained there three days without food or
shelter.   They might have found their way over the
hill, but a dense fog covered all the island.   At length
some women, happening to go to the glen, were seen
by the sailors, who shouted and pursued the women,
who ran off like hares.   The men, however, followed,
and were hospitably received, and got the loan of a
boat to take them to the mainland.   The old man
did not know the name of the ship, nor where she
came from, but the men all talked *Beurla*—English.
In an hour or so the five St Kildians came down the
hill again, and reported that although they had
shouted loudly, they had heard no reply, and they
were sure the girls were mistaken.   The latter, however, stoutly persisted, "*Tha goill air a gleann.*"   Some
women and children, who had been hiding in *clætyan*
at the back of the village, now ventured to go home.
Next morning a steamer was seen making away from
the island, and it was no doubt her fog-whistle that
had startled the village from its propriety.   I relate
this incident to show how easily the people of St
Kilda are excited, especially if the object of alarm is
connected with the mysterious world outside.

In October, when the nights were getting long,
spinning wheels began to be busy in every house

making the thread which the men afterwards wove into cloth; and I spent the evenings in one or other of the cottages, chatting with the people, and endeavouring to conquer the difficulties of Gaelic, and to penetrate into minds that had been trained in such peculiar circumstances. I tried, with many a blunder, no doubt, to tell them stories, such as Blue Beard, in which they seemed to feel a deep interest; the women sometimes improving my grammar, and helping me out of a difficulty. They would also tell me artless *sgeulachdan,* or tales, of which the following are samples. By comparing them with the popular tales of the Long Island and other places, some little light may be thrown upon the history of St Kilda.

A Fairy Tale.

"A long time ago, two men in passing a spot in the village where a house now stands, but which was then a little green hill (*cnoc gorm*), heard a sound as if some one were churning butter in an earthen jar. One of the men cried out, 'Give us a drink, goodwife,' when a door immediately opened in the hillock, and a strange woman came out, and presented a bowl of milk to the man who had asked for it; but he declining to take it, she handed it to the other man, who accepted it with thanks, and drank the milk. She then went into the hillock, and the door closed.

Mór Bhán (Matt xxv 35.36 v.)

That same day, strange to relate, the man who had refused the milk fell over the rocks of *Osimhal* and was killed. The other lived long and happily."

It is curious that in digging the foundation for a house at this little hill, an ancient structure, similar to *Airidh mhòr*, was discovered.

THE GIFT OF THE GAB.

" One day a St Kilda woman was sitting alone in a hut, rocking her child in a cradle, when two strange women, dressed in green, entered the door, and, by some magical power, deprived her of the power of speech, so that she could not call to her neighbours; but .she heard one of the women say to the other, ' This child, I see, has drunk of the milk of the cow that ate the *mothan*, and we can do nothing for him except give him the talent for language.' When the child grew into a man, it is related that he possessed an extraordinary fluency, could compose a rhyme on any subject on the shortest notice, and would talk more than any six men on the island—a questionable gift for his neighbours. His mother died in Harris when the grandfathers of old men were living."

A CHANGELING.

" It was in the harvest time, when a couple went into a croft in front of the village to pull their corn.

The woman carried an infant, and before beginning her work she walked up to a hillock to lay her child upon it. The man looked alarmed, and earnestly entreated her not to lay the infant in that place. 'Mind your own business,' said she. 'Take your own way then, as you always do,' he answered. In a short time she went to take the child off the hillock again, when her husband beseeched her, for the love of God, not to remove the child for a little. Happening to be in a more compliant temper, she did as she was asked; and it was a fortunate thing for her, because the man being gifted with the second sight, had seen the fairies come out of the hillock, take away the child, and leave an ugly goggle-eyed brat of their own in its place."

THE DEVIL IN THE SHAPE OF A BULL.

"Long, long ago, a party of St Kildians happened to be in the island of Boreray, and were living in the *Taigh an Stallir*, or the Hermit's House, and being short of provisions, one of them expressed the wish that they had the fattest ox in Clanranald's herd, when, on the instant, the lowing of an ox was heard outside. 'There,' said one of the party, 'now your wish has been granted; go out and kill him.' The man was too frightened to go out, but next morning the marks of cloven hoofs were to be seen on the mud outside."

This simple story is interesting from its reference to the name of Clanranald, who possessed the island at a distant although indefinite date, and from its thus showing how long these traditions may float down the stream of time.

A WATER BULL.

"One day a man who came down the *lag*, or hollow at the back of the village, with a burden of peats on his back, saw a door open in the side of a small conical hill. With great presence of mind he whipped out his knife, and stuck it in the ground at the foot of the door, and as he gazed, lo! a spotted bull came out and dropped a cow. This cow in course of time produced a calf which had no ears."*

There is a story about a man who shot a *tarbh uisge*, or water-bull, with a bow and arrow, in a lake on the top of *Sgal*.

THE WELL OF YOUTH.

"Once on a time an old fellow, in going up Connagher with a sheep on his back, observed a well which he had never seen or heard of before. The

---

* "There are numerous lakes where water-bulls are supposed to exist, and their progeny is supposed to be easily known by their short ears."—Campbell's Popular Tales of the West Highlands, Intro., p. xcvii.

water looked like cream, and was so tempting, that he knelt down and took a hearty drink. To his surprise all the infirmities of age immediately left him, and all the vigour and activity of youth returned. He laid down the sheep to mark the spot, and ran down the hill to tell his neighbours. But when he came up again neither sheep nor well were to be found, nor has any one been able to find the *Tobair na h-'oige* to this day. Some say that if he had left a small bit of iron at the well—a *brog* with a *tacket.* in it would have done quite well—the fairies would have been unable to take back their gift."

I made inquiries about Lady Grange, who was forcibly sent by her husband to St Kilda in 1734, and kept there for seven years. Her name was familiar to all the old people and to some of the young. Tradition says that she slept during the day, and got up at night,—" Sun, I hate thy beams." She never learned Gaelic. The house in which she lived belonged to the steward, and was a little larger than the old huts still standing, but similar otherwise, and was demolished a few years ago. A dearth happened to prevail during the whole time she resided on the island, but she got an ample share of what little food there was. The best turf was provided for her fire, and the spot where it was got is still called "*Poll na Ban-tighearna*," or the Lady's Pool. She was much beloved, and the people presented her with a straw chair, as a token of respect, when she was carried off to Harris. I heard nothing of her violent temper. Perhaps she had some reason to be violent when at home!

CHAPTER VIII.

N the 21st of October, and for many sub-sequent days, all the inhabitants went down the cliffs to pluck grass for their cattle, which, as the proprietor had failed to send a smack for the young beasts, were about double the number usually requiring to be fed. I saw the women lying on the narrow sloping ledges on the face of the rocks. A slip, or a false step, and they would have fallen into the sea hundreds of feet below, or been dashed to jelly on the projecting crags. I frequently went to the top of the hills, but seldom saw a sail even on the far horizon—never near the island.

On the 7th November a meeting was held in the church to return thanks for the miserable harvest. A sudden change occurred in the weather; the sky became charged with thick vapour, and there was a

Hill-Kop St. Kilda.

heavy fall of hail, accompanied by thunder and lightning.

On the 8th of December I went to the top of the hills, and notwithstanding my "lenten entertainment," felt remarkable well ; but slipping when about twenty yards from home, I sprained my ankle, and lay for some time in torture. I crawled into the house, and after a time succeeded in cooking my dinner, such as it was. I slept none, and next day my room was filled with sympathising male friends and ministering angels. Some brought me presents of salt mutton, potatoes, turf, and fulmar oil. On the 10th I held a levee, the whole people coming to see me between fore and afternoon services. The men about this time began to weave the thread which the women had spun. Both sexes worked from dawn of day until an hour or two after midnight. Their industry and endurance astonished me. I soon began to limp about ; and when the nights were dark, I got a live peat stuck on the end of a stick to let me see the road home. At this time I made a miniature ship, and put a letter in her hold, in the hope that she might reach some place where there was a post-office, being anxious to let my friends know that I was alive, and also to let the public know that MacLeod had broken his promise to send provisions to the people, and that we were all in want. Shortly afterwards I made a lanthorn out of a piece of copper that had come off a ship's bottom. A large limpet-shell filled with fulmar oil served for a lamp inside.

On the 12th of January, which is New Year's day in St Kilda, the minister, to celebrate the occasion, preached a sermon from that cheerful little book Ecclesiastes.

On the 17th the most remarkable event occurred that had happened in St Kilda for many years. The people had just gone to church, when, happening to look out at my door, I was astonished to see a boat in the bay. Scarcely believing my eye-sight, I ran down to the shore—past the church—and took a closer look. Eight or nine men, all in sou'-westers and oil-skins, sat resting on their oars in a white boat. Robinson Crusoe, when he saw the foot-print on the sand, could scarcely have felt more surprised than I did on beholding this apparition. I bawled as loudly as I could, but my voice was drowned by the roar of the waves. A woman who had followed me gave notice to the congregation, and all came pouring out of the church. The St Kildians ran round the rocks to a spot where there was less surf, and I went with them, and waved on the boat to follow. When we arrived at the place mentioned, the islanders threw ropes from the cliffs to the men in the boat, and I shouted, "Tie the rope around your waist, and the men will pull you up;" when a voice replied, "Oh, tank you! mooch better dere." It was the captain who spoke, whilst pointing to the flat shore in front of the village and putting the boat about. All ran back, but before we had reached the shore, the strange boat had run through the surf. Instantly on touching the sunken rocks, all the men leapt into the sea and swam to the land, where the St Kildians were ready to grasp them. In a few minutes their boat was knocked to pieces on the rocks, and they were prisoners like myself.

The strangers were invited into the minister's house, and dry clothes given them. When the captain was able to speak, he told us that he had left Glasgow for New York, on board of the Austrian ship *Peti Dabrovacki*, 880 tons, five days before; that the ship was in ballast, which had shifted in consequence of the bad weather she had encountered, and she had fallen on her beam ends and become perfectly unmanageable; that he and eight of the crew had left her, eight miles to the westward, in the boat, and that seven men refused to leave the ship, and would probably perish—as no doubt they did— for the ship was not to be seen next day. When he had answered my interrogations, he put the following questions to me :·

He.—"Vat island is dis?"

I.—"St Kilda."

He.—". Is dere a telegraph?"

I.—"Telegraph! God bless you! No."

He.—"Is dere a post?"

I.—"No! no post."

He.—"Is dere a steamer?"

I.—"A steamer sometimes calls by chance, once in two or three years, perhaps. A small smack called here last June, and ought to have called again in August, but she did not come." ·

He.—"Who does dis wretched island belong to?"

I.—"To Great Britain, I believe," and I blushed for my country.

He.—"How long have you been here?"

I.—"I have been a prisoner here for nearly seven months."

He.—"A prisoner!" (and here he glances sharply from my brogues to my bonnet, fancying, as he afterwards told me, that he had got into a penal settlement). "Oh dear! dear! vot am I to do. I have one wife and one old mother; they will think I am dead, and put on black cloth."

I.—"I am sorry for you, but there is no help."

He.—"But I have no tobacco, and I have smoked for forty year."

I.—"Chew your waistcoat pocket, as we have been

all doing, and wait patiently until Spring, when it will be possible for a boat to reach Harris."

When the strangers had shifted their clothes they were distributed amongst the sixteen families that compose the community, the minister keeping the captain, and every two families taking charge of one man, and providing him with a bed and board and clean clothes. I myself saw one man (Tormad Gillies) take a new jacket out of the box into which it had been carefully folded, and, with a look of genuine pity, give it to the mate to wear during his stay, as the young man sat shivering in an oil-skin. The oatmeal being done, the islanders took the grain they had set apart for seed, and ground it into meal for the shipwrecked men. They put up bunks in the rooms of married couples for the strangers to sleep in, and several of the natives asked me to look if the beds were good enough. This hospitable conduct of the St Kildians is all the more commendable when one considers that their guests were all foreigners and Papists, whom they had been taught to hold in horror. But long before the five weeks had elapsed during which the Austrians lived on the island, they had, by their good conduct, removed the prejudice that had prevailed against them at first. Most of them knew a little English. It is remarkable that all the natives, without excep-

H

tion, caught the *cnatan na gall*, or strangers' cold, on the arrival of the smack last year, and that every one again suffered from the same illness on the arrival of these castaways this year. The natives firmly believe, as their ancestors did, that they are certain to catch cold whenever a boat or ship visits the island. This illness is sometimes called the boat-cold and sometimes the Harris cold. They say it is not so severe when a yacht comes from Glasgow or Liverpool. They have been laughed at for this belief, and I have joined in the laugh, but after all there may be some truth in it. It may be possible that they are more apt to catch disease than those who are accustomed to an infected atmosphere. The *cnatan na gall* resembles influenza.

On the 28th January the wind blew violently from the north-west, with heavy showers of sleet. The huge waves came rolling into the bay against the wind, which caught them as they fell on the shore, and carried them off in spin drift. The air was so full of sleet and spray that the Dun was invisible. It was the worst day I had seen in St Kilda, and yet many of the women hurried past to church barefoot.

There are two services and a Bible-class, to all of which everybody attends every Sunday. They occupy altogether six hours and a half. The church in Winter is a miserable place. Mould adheres to the

The Minister.

walls like hoar frost and feathers, the benches, many of them bearing the mark of the ship-worm, are without backs, and there is no flooring on the ground.

The best resident ruler, "guide, philosopher, and friend," for St Kilda would be a sensible, firm, and good-tempered old sailor, able to work and repair a boat, to teach the three R's and a little English to the young, and to scrape a reel on the fiddle for the girls to dance to; and the worst home ruler would be a well-meaning but feeble-minded, irresolute, yet domineering fanatic, whose servant would lead him by the nose, and get him to preach at any woman to whom she had a spite, who would be obliged to sit and listen in silence, however innocent. This latter character is, of course, entirely supposititious, but it is quite possible that the Free Church might send such a representative to St Kilda, to sit like an incubus on the breast of the community. In that sequestered island, beyond the supervision of Sessions and Presbyteries, he might, by working on the religious prejudices of his flock, retain his grasp, and exercise a tyranny which would never be tolerated in other places.

Those who believe that Tenterden Steeple was the cause of the Goodwin Sands may attribute the virtues of the St Kildians to the influence of the ministers; but the islanders manifested the same good qualities before a clergyman lived amongst them. They were distinguished for their kindness to strangers, by their affection for each other, by their decency and sobriety, by their piety and industry, two centuries ago. "They that are whole need not a physician; but they that are sick."

On the 29th the captain and sailors called on me and felt interested in seeing a canoe I had hewn out of a log. I had written a letter and put it into her hold, enclosed in a pickle bottle. The captain begged me to write a note for him, addressed to the Austrian Consul, Glasgow, and to put it in another bottle. The sailors, glad of anything in the shape of work, helped me to rig her and put the iron ballast right, and to caulk the deck. We delayed launching her until the wind should blow from the N.W., which we hoped would carry her to Uist or other place where there is a post. A small sail was put on her, and with a hot iron I printed on her deck, "*Open this.*"

The sailors at the same time carried the fragments of their boat to the top of *Sgal* and *Osimhal*, and made fires at night, in the hope that the people in the Long Island might see them and send a smack to their relief.

The captain brought me a life-buoy belonging to the lost ship, and said he intended to send it off. I suggested that another bottle should be tied to it, with a note enclosed to the Austrian Consul, and that a small sail should be erected. This was done, and the life-buoy was thrown into the sea, and went away slowly before the wind. None of us had much hope that this circular vessel would be of service. She was despatched on the 30th, and, strange to say, reached Birsay in Orkney, and was forwarded to Lloyd's agent in Stromness on the 8th February, having performed the passage in nine days. During my residence in St Kilda, several canes, one of them hollow and several inches in diameter, which the Gulf Stream had brought from some tropical clime, were picked up by the men, who split the canes they find to make or repair the reeds of their looms. I had not calculated on the strength of this current; but "there's a Divinity that shapes our ends, rough-hew them how we will."

On the 5th of February we sent off the canoe, the wind being in the N.W., and continuing so for some days. She went to Poolewe in Ross-shire, where she was found lying on a sand-bank on the 27th by a Mr John Mackenzie, who posted the letters.

On the 17th February, the Austrian skipper (who was in a very desponding state) offered L.10 for a passage to Harris in the new boat for himself and men. The St Kildians firmly declined to let him have the boat, unless a crew of their own went with him; and it was finally agreed that seven natives and myself should go in her. The St Kildians, knowing the dangers of the passage at that season, drew lots who were to go. The captain said that seventeen men would be too many for the boat's capacity, and hinted that I had better remain in St Kilda; but the natives insisted on my going, in order that I might represent the condition of the island to the public, and try and get provisions sent. The Ross-shire woman (who was indebted to my good word for a clock, and the largest stock of carpenters tools ever seen or heard of in the island, besides many gifts out of my own pocket) suggested that I should be made to pay for my passage in the boat I had been the means of getting for them,—a degree of ingratitude with which I was highly amused. The men, however, would not listen to this; and all was settled but the weather. We were waiting for a promising day, when,

On the 22nd, about seven in the morning "by most of the clocks," as I was lying awake in bed, and thinking upon getting up to make my gruel, I was startled by hearing the shriek of a steam whistle. I sat bolt-upright, and then lay down again, muttering "It was the wind," when hark! the whistle is repeated. I leaped out of the straw, ran to the door, and saw,

sure enough, a steamer in the bay. Seizing the first
article of clothing that came to hand, and pulling
them on the wrong way, I rushed barefoot out of the
house, and up the street, rattling at every door, and
shouting "*Goill!* steamer! *Erich! greas ort!* In a
few minutes all the people were astir and hurrying
to the shore. I went back and finished my toilet,
packed my trunk, and running to the shore stepped
into a boat that I saw belonged to Her Majesty.
Then I discovered that I had left my purse and
other property in the house; but the sea was rising,
and the surf was too great to allow me to land. I
had received a mandate, written in Gaelic by the
minister, and signed by him and by all the men who
could write, authorising me to do my utmost to get a
mail sent to the island twice a year. The men had
also requested me to try and get provisions and seed-
corn sent to them. I undertook the business; but
the minister, now jealous of his authority as dictator,
pushed on board of the steamer (which I saw was
H.M.S. *Jackal*), and explained the state of the island
to the commander, who telegraphed the fact to the
Admiralty, who took no notice of it. The captain,
however, left some biscuits and some oatmeal,—a
three week's supply,—and I saw his reverence
hobbling out of the vessel with three pounds of
tobacco in his hand. "How did you know we were

here?" I asked one of the officers. "Why, your letter came on shore at Birsay, and the Admiralty telegraphed that we were to take in coal, and to proceed to the relief of the Austrians immediately." I stepped to the side of the vessel, and cried out to the St Kildians who were alongside, "It was the lifebuoy brought the ship here, you unbelieving fellows!" for they had laughed, although in a good-natured compassionate way, at my desperate efforts to get a letter sent to the mainland. "*Bha!*" they answered. Waving an adieu to my kind friends in the boat and on shore, we were speedily under weigh. "By Jove! did you notice," said the officer, "what an active hand these Austrian fellows lent in getting up the anchor?"

The shipwrecked skipper and I were accommodated aft, and I found all the officers like friends and brothers, and here return my thanks to them. I was made to feel quite at home. The Austrians were treated with equal hospitality forward, and seemed particularly delighted at again having as much tobacco as they could use, for we had been all smoking dried moss for some time. I got myself weighed, and found I had lost about 30 lbs. of superfluous flesh during my sojourn of eight months in St Kilda.

The wind had risen, and the sea became rough

and if the *Jackal* had been half-an-hour later she would have been obliged to return with her errand unexecuted, for it would have been impossible for a boat to approach the shore. We reached Harris in the afternoon, and anchored in the Sound all night.

We arrived at Oban next day, where I was interviewed, and the destitute condition of St Kilda published all over the kingdom. We arrived at Greenock on the 26th, where a crowd of newspaper reporters were waiting our arrival, and to whom I communicated a full, true, and particular account of my eight months' experience, in the belief that the more light that was thrown upon the condition of St Kilda, the greater was the hope of its being ameliorated. I reached home penniless and all but barefoot, but in good health and spirits.

CHAPTER IX.

My mission—The extraordinary way in which Providence has assisted me thus far, which inspires me with hope for the future—A nine days' visit from Fame—The Kelso Fund— The Treasurer hangs fire, but goes off with a double charge at the end of two months — Unsuccessful attempt to fly a kite—What should be done with the rump of the Kelso Fund.

TO break open the door of MacLeod's prison was the object of my second visit to St Kilda. To liberate the poor serfs who had been so long incarcerated and cruelly used, and to bring them into communication with the rest of the world, was my mission, and often, when rambling amongst the stern rocks on the tops of the mountains, or sitting listening to the solemn sound of the waves upon the lonely shore, I felt as if I had had a Divine call to perform the work, and must proceed at any cost, and despite of any opposition. Providence often selects strange instruments with which to execute His purposes,—instruments that would seem altogether unsuitable to Doctors Begg and M'Lauchlan.

I found that the boat I had taken out would not accomplish the end I had in view, and I became convinced that no private boat would answer. I saw that the men of St Kilda, however much they might grumble and pant for freedom, had become, from their life-long imprisonment, like children in the dark, and were frightened at shadows, and that the minister (whose prayer was, "Give us peace in *our* time, O Lord!") used his influence to paralyse what little enterprise the people possessed. I felt convinced, moreover, that even if the islanders had the resolution to go to Harris for purposes of commerce, the proprietor would not allow them; that he was determined to retain his monopoly and to resist all reform. In a threatening letter he had written to his poor tenants in October 1875, he tells them he is opposed to all change, and if they continue to grumble, he will let the island to one man—a middleman—the St Kildians having had bitter experience of the tender mercies of that class. In a letter written subsequently, but delivered at the same time, and after I had exposed the condition of the island in the first edition of this book and in the newspapers, MacLeod seemed to come down a peg, and agreed to let the people go to Harris and conduct their own business; but it was evident from the tenor that he meant

> "To keep the word of promise to the ear,
> And break it to the hope."

I accordingly came to the conclusion that nothing but a steamer would answer the purpose I had in view, and I thought that the St Kildians, being British subjects and paying taxes, had surely a right to some little postal communication—say even twice a year. They all agreed that this would be an inexpressible boon, and begged me to try and get a steamer sent to the island in Spring and Autumn, and gave me an *ughdarus*, or mandate, to act as their *fear-ionad*, or representative.

There is a Providence in the fall of a sparrow, and every event that occurred in St Kilda during my sojourn of eight months, although it appeared like a disaster at the time, seemed in the end as if it had been specially appointed to further the cause I had undertaken to advocate. Such a concurrence of remarkable events had never before happened in the island, and it seemed as if I had been predestined to be detained to witness and publish them. Firstly, Donald MacKinnon, who was supposed to have been drowned thirteen years before, arose again, as it were, from the dead, and hopes were excited that others of the lost crew might have been preserved

also. I was there to see the mental agony, the
suspense, the alternate hopes and fears, which the
people had to endure for eight weary months, and
to realise to the full what an atrocious neglect the
Government was guilty of in leaving these seventy-
five poor British subjects upon that rock without a
mail. One poor old fellow would grasp my arm, and
entreat me, whilst the tears stood in his eyes, to
send him news when I got to Harris about his lost
son. His voice, tremulous with emotion, still sounds
in my ear,—" *Mo h-aon mhac ! mo h-aon mhac !*—My
only son ! my only son !" and my heart boils with
pity for these poor people, and with unutterable
hatred for the cold-hearted wretches who try to keep
them thus in darkness and in prison. Secondly,
I seem to have been detained on the island to witness
the breach of contract of which MacLeod was guilty
—to hear the factor promise to send oatmeal in
Autumn, and to see the sufferings the people endured
by his failure to do so. "But they could not have
suffered much in St Kilda," said a servant of Jesus
Christ ; "they must have had fulmars and they had
sheep." The St Kildians are not fastidious in their
tastes, and when they say they are in want of food they
may be believed. The supply of fulmars is limited,
and the sheep in Winter are mere skin and bone. I
could not peep into every pickle tub in the village to

see how many fulmars were left, and I never doubted the truth of what the people said. "The young may see the Spring," I heard one man remark, "but the old will die." This prophecy may not have been literally fulfilled ; but I saw every one turning leaner, and some of delicate constitutions were seized by various diseases, which had evidently been induced by low living. Those still suffer, and will probably die. One is a young lad called *Ian Bàn,* who was the support and comfort of his widowed mother. Mac-Leod's breach of contract on such a serious matter was little short of culpable homicide. Thirdly, I was present to witness the arrival of the Austrian casta-ways. The addition of nine hungry stomachs to the population in a season of dearth seemed at the time to be a great misfortune; yet it led to my escape from captivity, and to getting relief sent to St Kilda. It also afforded an excellent test of the religious principles of my constituents, and proved that their Christianity was not all theoretical, as it is too fre-quently the case in more civilised places, but was a light and a guide to their worldly conduct. They had not only studied the parable of the good Samari-tan, but they copied him ; and in the reception they gave these poor destitute foreigners, they even im-proved on the model. For they ate the plainest fare themselves, and gave the best they had, and without stint, to the unfortunate strangers (*goill agus papinich, ach créutairan bochd*); grinding, in their reckless hospitality, even the very seed-corn, on which, so far as they knew, their future subsistence depended. No shipwrecked sailors had landed on the island for thirteen years, and it seems to me as if Providence had ordained that I should be present to witness the noble conduct of my friends, so that I might publish it to the world, and excite sympathy on their behalf. Fourthly, the marvellous manner in which my letter was conveyed to Orkney seems providential. The chances were that it would drift away into some unvisited part of the ocean, or fall into the hands of some callous, incredulous, or illiterate person who would not post it.

And when I arrive at home, I find (thanks to the press) that a full account of all the occurrences above specified has been circulated throughout the kingdom, and that a deep interest is felt in St Kilda, and that the public is ready to listen to my plain, unvarnished tale. If I had had an opportunity of leaving the island in Summer, I would have tried to advocate the claims of my constituents; but I would have had to roar into deaf and inattentive ears. I am not at all of a pious turn, but I firmly believe that there are "more things in heaven and earth than are dreamt of in our philosophy."

Fame is a lady who looks best at a distance, and
the wild wail of the bagpipe in some solitary glen is
sweeter to me than the discordant bray of her old
trumpet heard in the crowded city.   She is a treach-
erous jade, and if a man is so foolish as to love her
for herself, she is certain to lead him into trouble,
flattering and caressing him for a day, and then
tearing his clothes to tatters and biting him with her
serpent-fangs, or decoying him into some exposed
place, where the rabble can hide behind hedges and
pelt him with filth, until those who pity him most are
afraid to approach him for the smell. Some of the,
best of fellows have been thus beguiled and destroyed
by the strumpet.   I had a visit from this dangerous
person on my return home, and she remained with
me nine days.   I received her with civility, believing
that she might be of service in forwarding the business
I had in hand ; but I kept a suspicious eye on her,
and yet, with all my circumspection, the jade
managed, during the brief period we spent together,
to play me a dirty trick or two, and to give me some
trouble.

To drop allegory.   Letters by the dozen reached me
nearly every post from all parts of Great Britain.
Some were from people of genuine benevolence, who
tendered their services in collecting subscriptions for
the relief of the St Kildians.   Others were from

persons who had an eye to the main chance, or had questions to put, or suggestions to offer; and I was overwhelmed with correspondence. One individual reprobates the extortionate profits exacted by Mac-Leod for groceries, and offers to supply sugar and soap, of the best quality, at twopence a pound less than what was charged by that aristocratic shop-keeper. Another hates oppression, and expresses the deepest sympathy with the inhabitants of St Kilda, and places a steamer at my service, so that provisions and seed-corn may be forwarded immediately—if I will undertake to get up a party of forty passengers at L.5 a head, and pay for coals. A firm of cattle-dealers applaud my efforts for the amelioration of St Kilda, and hope that I will introduce a small breed of cows for which they are agents. A Celt requests me to write and tell him the surnames of the St Kildians, and whether there are any MacCraws on the island. One sagacious individual expatiates on the fecundity of rabbits, and suggests that a few pairs should be let loose, so as to supplant the fulmars. An answer will oblige. I thought of sending a batch of these letters to MacLeod, in the hope that they might amuse him; but remembering the dog-in-the-manger aver-sion he has to any one interfering in the management of his rocky estate, I refrained for fear the perusal might throw him into a fit of apoplexy.

I

To get immediate relief sent out to my distressed
constituents, and afterwards to get a mail steamer to
call twice a year at their island, were the principal
tasks I had set myself to accomplish.   Just as I had
resolved to start a subscription in Edinburgh, and to
accept the offers of assistance that had come from
Kelso, Perth, and other towns, I was informed that a
fund existed for the special benefit of the people of
St Kilda in seasons of distress, and that Mr Menzies,
Secretary of the Highland Society, was the treasurer.
I called on that gentleman on the 2nd March, and
was told that the money, which originally amounted
to L.600, had been left by a benevolent gentleman
called Kelso, who had died in the West Indies; that
L.254 of it had been spent at various times in the
purchase of a bull, a boat, seed-corn, etc.   The
management of the fund, he explained, had been
left at his individual discretion, and his intromissions
were not entered in the accounts of the Society.   He
mentioned that it had been considered prudent not
to enlighten the people of St Kilda as to the existence
of the fund, lest they might have depended upon it
and been spoiled.   I replied that I should consider
it my duty to tell them all I had learned on the
subject, so that they might know to whom they were
indebted for the assistance they had received.   They
fancied that the boat and bull had been presented

to them by the former proprietor. Mr Menzies solemnly promised that L.100 should be advanced out of the fund to purchase seed-corn and provisions, and that Government would be applied to for the use of a steamer to carry them out. He would attend to the business immediately, and I need give myself no further concern on the subject.

Relying on this promise, I declined the offers of assistance that had been made me by gentlemen in different parts of the country.

On the day after my interview with Mr Menzies, an anonymous attack on me appeared in the *Scotsman*, and I was charged with ungratefully neglecting the interests of the humble friends who had been so kind to me. This letter was evidently a little bit of sharp practice on the part of an Edinburgh lawyer with a view to business, and to pave the way for a meeting, which was held on the 5th, and to which I received an invitation at the eleventh hour. I was unable to attend, and now that immediate relief had been promised, I saw no necessity for an appeal to the public. At the meeting I was accused of having concocted a got-up story, of telling false-hoods, and of uttering exaggerations and mis-state-ments. I tried as well as I could to defend myself with my pen. If I had been inclined to exaggerate the state of destitution in St Kilda, it would surely have been when I was a prisoner there, and anxious to get away, and not when I had arrived at home, and had no selfish motive for mendicity; and yet in my letter sent by the buoy I merely said that "provisions are scarce." All that I said regarding the condition of St Kilda has been subsequently verified by the actions and reluctant and indirect admissions of my opponents, not one of whom, however, has had the courage or the conscience to say so directly, or to add that he was sorry he had doubted my word.

A Committee arose out of this meeting, like a kite with my name tied on for a weight to the tail. Indignant at the unjust treatment I had received from some of the members, I cut the connection and let the glittering machine ascend without me. The gentleman who had taken the greatest pains to make it fly still kept running with the string in his hand against the wind, with triumphant huzzas, too; but all his efforts could not keep the kite aloft. After a few oscillations it took a dive to the earth, never to mount again, to the amusement of the public, who, notwithstanding the most urgent appeals, would scarcely put enough into the hat to defray expenses.

Meanwhile a month elapses, and no provision or seed-corn is sent out to St Kilda—a month is a long time for hungry people to wait for food. The Secretary of the Committee announces that H.M.S. *Jackal* has just left with supplies—a statement which is entirely without foundation, and which shakes the faith of the public in the Committee. I receive a letter from Mr Menzies, saying that the Admiralty had refused the use of the *Jackal* until MacLeod had made an attempt to land the stores which ought to have been delivered eight months before. I get alarmed at the delay, and proceed to Greenock to raise subscriptions to purchase provisions, and to charter a ship to carry them out immediately. This was about the 24th of March. People seem sympathetic; but just as I have begun to receive contributions, I learn that the proprietor's smack has sailed from Dunvegan with supplies, and I return the money I had received and go home again. The smack, after a long delay, succeeds in reaching the island, and brings back the intelligence that the " supplies were much required," and that the people had been " abstaining for some time from their evening meal," which, put into plain language, means that they had no dinner—nothing between ten A.M. and ten P.M., and little enough then, no doubt. Probably Mr Menzies had received an answer from the minister, to whom he had written, confirming all that I had said as to the condition of the island, but he did not publish the fact, because it would have cleared my character from the aspersions that had been cast on it, and would have been a reflection on his own dilatoriness. But he sends out supplies by H.M.S. *Flirt* (upwards of two months after he had been apprised of the destitution of St Kilda), and not only expended L.70 of the L.100 he had promised to disburse out of the Kelso Fund, but also spent L.100 which the Austrian Government had presented to the people of St Kilda, in the purchase of provisions and agricultural implements, thus, by his actions, more than confirming the truth of my statements.

The Kelso Fund was left by the benevolent donor to assist the people of St Kilda in seasons of emergency. But when distress occurs, how are the beneficiaries to let the treasurer know of it? or if a messenger, by a rare chance, happens to get from St Kilda to Edinburgh, how is he to make Mr Menzies believe that he is speaking the truth?

The best plan would be to keep a store of bread stuffs in the island, to be paid for out of this fund. Such a store might be safely left in charge of the people, with the injunction, communicated to all, that it was not to be used except in a case of urgent necessity. The money contributed at present to the Free Church should be devoted to the same purpose. The Kelso Fund, if it had been wisely and economi-

Donull Og.

cally managed, might have been of great and per-manent service to the islanders ; but a large portion seems to have been squandered in the purchase of useless implements, etc. The interest on the original sum (if the principal was invested, and not kept in a ram's horn) would have been more than sufficient to meet an emergency.

Part of the interest on the balance might be employed in assisting the poor, to whom an annual dole of oatmeal would be a great boon. I may mention the names of the most necessitous persons in St Kilda :

1. Donald MacQueen, generally called Donull Og, who is an old man, crippled with rupture. He has an old wife and daughter. He is assisted by his son, who is a married man, but the total earnings of himself and family for the year 1875 only amounted to L.7, 5s. 8d., for which he would receive goods on which a large profit is charged. He is in debt to the factor, and, in his ineffectual efforts to get it cleared off, reminds one of Sisyphus.

2. Rachael MacCrimmon, a middle-aged unmarried woman of estimable character. She lives all alone in one of the old huts, and has a few sheep, from the wool of which, made into cloth, and from the feathers of the birds she catches, she derived an income of

L.1, 19s. 7d. in the year 1875. As she has no land, what oatmeal she consumed must have been purchased out of this sum. Having no cow, she is obliged in general to take her *brochan* without milk. The only luxury she indulges in is three ounces of lozenges per annum, which cost her a sixpence. She is desolate and destitute, but bears her hard lot with Christian resignation.

3. *Mòr-nighean-Mhic-Ruairidh*, the deformed and delicate daughter of two poor old people.

4. Ruairidh MacIan, an imbecile, upwards of sixty years of age, who lives in an old hut by himself.

In conclusion, I may mention that seven vessels (five of them steamers) have to my knowledge called at St Kilda this Summer, although the weather has been anything but propitious. So many vessels never called in one season at that island before, and the islanders express astonishment at the number of strangers that have called to see them. I have received five letters written in Gaelic from three of the natives—the first they have ever written. This will show them that correspondence can be conducted in their mother tongue as well as in English, and will induce them to practice writing. Attempts have been made by some of the visitors to blacken my character in St Kilda with baseless falsehoods; but I flatter myself the people know me too well to believe such rumours. But could my enemies succeed in making me as black as the ace of spades, it would not help their cause a bit, or make any person of sense or feeling believe that the

TRUCK-AND-NON-POSTAL SYSTEM

is the correct thing. Like rotten bones it will drop to dust on being exposed to the light. I flatter myself that I have kindled a flame which MacLeod and his friends will find some difficulty in extinguishing, no matter how many wet blankets and petticoats they may use:

"Facts are chiels that winna ding."

APPENDIX.

MANDATE FROM ST KILDIANS.

Iorta, an seachdamh la de cheud mhios an Fhoghair 's a' bhliadhna 1876.—Tha sinne luchd-aiteachaidh Eilein Iorta a' guidhe a bhi 'tabhairt fios do na h-uile a ghabh gnothuch ris a ni so, gu'n d'iarr sinn air Maighistir Sands 'n uair a bha e so, 's a' bhliadhna a chaidh seachad, gu'm faigheadh e eithear dhuinn, ni chaidh leis gu h-iongantach le cuideachadh chairdean, agus gu 'n do liubhaireadh leis dhuinn i gu tearuinte. Tha sinn mar an ceudna, 'guidhe a bhi cur 'an ceill gu bheil sinn buileach toilichte leis an eithear, agus gu bheil i gle fhreagarrach air son gnothuich an Eilein aig an àm so, agus tha sinn a' tabhairt mòr bhuidheachais do ar luchd-cuideachaidh air son an tiodhlaic mhòir a chuir iad thugainn. Air dhuinn fhoghlum gu bheil sanas air a thabhairt seachad gu'm bu choir do luing gairm aig an Eilean so, da uair 's a' bhliadhna, chum sinne thoirt gu comh-chainnt re ar

comh-chreutairean, tha sinn leis so, ag orduchadh
agus a' tabhairt ughdarrais do Mhaighistir Sands mar
ar fear-ionaid, gu'n deanadh e na tha na 'chomas, gu
leithid so a shochair fhaotainn dhuinn. Na'n gair-
meadh aon de longan na smùid an so, da uair 's a'
bhliadhna, eadhon 's a' Ghiblean, agus am mios
deireannach an Fhoghair, bhiodh e na 'shochair agus
na 'ordugh ro fhreagarrach ar air son-ne, agus
chuidicheadh e gu ro mhòr ar sonas agus soirbh-
eachadh saoghalta. Agus tha sinn a' cur ar lamh-
sgriobhaidh ri so mar dhearbhadh air ar durachd agus
ar lan earbsadh ri eisdeachd fhaghail.

> John Mackay, Minister ; Neil M'Kinnon;
> Angus Gillies; Norman Gillies; Donald
> M'Queen ; Neill M'Donald ; Donald
> Ferguson ; Donald M'Donald ; Malcolm
> M'Kinnon.

I certify that the rest of the inhabitants concur in
the above, but, for want of practice, feel diffident in
signing their names.   JOHN M'KAY, Minister.

ENGLISH TRANSLATION OF THE SAME.

Hirta, the seventh day of the first month of
Autumn, in the year 1876.—We, the inhabitants of
the island of Hirta, pray to make known to all who

have concerned themselves in this matter, that we requested Mr Sands, when here last year, that he should get a boat for us, in which mission (with the assistance of his friends) he has been wonderfully successful, and that he has safely delivered it to us. We also desire to say that we are greatly pleased with the boat, and that it is very suitable for the business of the island at this time; and we gratefully thank those who assisted us for the great gift they have sent to us. Understanding that it has been suggested that a ship should call at this island twice during the year, that so we might be able to hold communication with our fellow-creatures, we hereby request and authorise Mr Sands, as our representative, that he do what lies in his power to obtain for us that convenience. If a steamer would call here twice a year, namely, in April and October, it would be an advantage, and a very good arrangement for us, and it would greatly add to our worldly happiness and prosperity; and we are putting our signatures to this as a proof of our wishes, and of our full confidence in obtaining a hearing.

(Signed) John Mackay, Minister; Neil M'Kinnon; Angus Gillies; Norman Gillies; Donald M'Queen; Neill M'Donald; Donald Ferguson; Donald M'Donald; Malcolm M'Kinnon.

I certify that the rest of the inhabitants concur in the above, but, for want of practice, feel diffident in signing their names.

(Signed)   JOHN M'KAY, Minister.

NOTE I.—In performance of the duty imposed upon me by the foregoing mandate, and stimulated by the hope that one man, thoroughly in earnest, might accomplish what a committee, composed of lukewarm and obstructive members, had pronounced to be impracticable, I have endeavoured, since my return home, to get the Government to send one of the steamers engaged in watching the Hebridean fisheries twice a year to St Kilda. Lord Elcho (after I had made several fruitless applications to others) had the goodness to forward my appeal to the Secretary of the Prime Minister, who will, no doubt, get his principal to take the matter into his *serious* considera-tion. When the Government walked by on the other side, I cried to the Board of Northern Lights to have compassion on my poor constituents, who had been robbed and maltreated for centuries, and left half dead. But the Board seems inclined to follow the Levite. A steamer belonging to the Board is in the custom of calling at the lighthouse on the island of Hysker once a month, and as that island is only thirty-two miles from St Kilda, I was in hopes that

the Board might have the humanity to order the
steamer to call at the latter island twice a year. My
petition has not yet been laid before the Board, but
the Secretary (who seems to be invested with extraordinary powers) favours me with his opinion that
it is " extremely improbable the Commissioners will
be able to comply with the request contained in your
application." Should I fail to get a mail for St
Kilda, I shall endeavour to persuade my constituents
to leave their prison, and go to some land where they
can enjoy freedom and justice and mercy.

NOTE II.—To the list of exports on page 59 add
twenty head of year-old cattle. The list is less the
contributions of Neill Fergusson, ground-officer.

H. & J. Pillans, Printers, 12 Thistle Street, Edinburgh.

MACLACHLAN & STEWART'S
LIST OF PUBLICATIONS.

Saxby, Henry L., M.D.
The Birds of Shetland, with Observations on their Habits, Migration, and Occasional Appearance. Edited by his Brother, Rev. STEPHEN H. SAXBY. 8vo, cloth, 21s.   [1874.

Gael, The.
An Gaidheal; A Monthly Magazine, devoted to Miscellaneous Gaelic Literature, and to the Interests of Scottish Highlanders Generally. Price 6d.; 5 vols., 8vo, cloth, 7s. 6d. each. Cloth Cases for binding The Gael, 8d.

Jerram, C. S.
Dàn an Deirg Agus Tiomna Ghuill (Dargo and Gaul); Two Poems from Dr SMITH's Collection, entitled the Sean Dàna. Newly Translated, with a Revised Gaelic Text, Notes, and Introduction, by C. S. JERRAM, M.A. 12mo, cloth, 2s. 6d.
                                                                 [1874.

Lismore's, Dean of, Book.
A Selection of Ancient Gaelic Poetry, from a Manuscript Collection made by Sir JAMES M'GREGOR, Dean of Lismore, in the beginning of the Sixteenth Century. Edited, with a Translation and Notes, by Rev. Dr M'LAUCHLAN, and an Introduction and Additional Notes by WM. F. SKENE. 8vo, cloth, 7s. 6d.   [1862.

Logan, James.
The Scottish Gael; or, Celtic Manners as Preserved among the Highlanders. Second Edition. Edited, with Memoir and Notes, by the Rev. ALEX. STEWART. 2 vols., 8vo, cloth, 28s.
                                                                 [1876.

M'Alpine, Neil.
A Pronouncing Gaelic-English and English-Gaelic Dictionary; to which is prefixed a Concise but most Comprehensive Gaelic Grammar. Seventh Edition, 12mo, cloth, 9s. [1877.
The above may also be had separately in 2 vols., 5s. each.

Rudiments of Gaelic Grammar. Eleventh Edition, 12mo, cloth, 1s. 6d.   [1872.

Mackenzie, Rev. Angus.
History of Scotland (Eachdraidh na H-Alba). 12mo, cloth, 3s. 6d.   [1867.

M'Lauchlan, Rev. Thomas, D.D.
Celtic Gleanings; or, Notices of the History and Literature of the Scottish Gael. 12mo, cloth, 2s. 6d.   [1857.

Mackenzie, John.
Sar-Obair Nam Bard Gaelach ; or, The Beauties of Gaelic
Poetry and Lives of the Highland Bards, with an Historical
Introduction by JAMES LOGAN. Fourth Edition, 8vo, cloth,
12s. [1877.

Munro, James.
A New Gaelic Primer, containing Elements of Pronuncia-
tion ; An Abridged Grammar ; Formation of Words ; also a
Copious Vocabulary. Fourth Edition, 12mo, cloth, 2s. [1873.

Ossian's Poems.
Dàna Oisein Mhic Fhinn, with a Preface by the Rev. Dr
M'LAUCHLAN. 12mo, cloth, 3s. [1859.

Stewart, Rev. Alexander.
Elements of Gaelic Grammar, in Four Parts. Third
Edition. Revised, with Preface, by the Rev. Dr M'LAUCHLAN.
12mo, cloth, 3s. 6d. [1876.

Bonar, Rev. Andrew R.
The Poets and Poetry of Scotland, from James I. to the
present time. Second Edition, 12mo, cloth, 3s. [1866.

Cumming, Rev. Jas. Elder, Minister of Sandyford
Church, Glasgow.
Abba Father ; being Ten Lectures on the Lord's Prayer.
12mo, cloth, 2s. 6d. [1862.

The Communicant's Manual ; a Simple and Practical
Guide to the Lord's Table. Third Edition, 18mo, sewed, 2d.
[1877.

Fraser, Rev. Robert W.
Elements of Physical Science ; or, Natural Philosophy in
the Form of a Narrative. Third Edition, 12mo, cloth, 2s. 6d.
[1858.

By the Author of " Mistura Curiosa."
Alter Ejusdem ; being another Instalment of Lilts and
Lyrics, with 150 Pen and Ink Sketches, and Occasional Music.
Second Edition, 25s. [1877.

M.P.
The Princess Iva ; A Christmas Story for the Children.
18mo, cloth, 1s. [1876.

MACLACHLAN & STEWART,
Booksellers to the University,
64 SOUTH BRIDGE, EDINBURGH.
(Opposite the University.)

Lightning Source UK Ltd.
Milton Keynes UK
UKHW022000250123
415976UK00005B/72

9 781016 811361